BrightRED Revision

Higher GEOGRAPHY

John Rutter

BrightRED
PUBLISHING

First published in 2009 by:

Bright Red Publishing Ltd
6 Stafford Street
Edinburgh
EH3 7AU

A CIP record for this book is available from the British Library

ISBN 978-1-906736-22-4

With thanks to Ken Vail Graphic Design, Cambridge (layout) and Roda Morrison (copy-edit)

Cover design by Caleb Rutherford – eidetic

Illustrations by Graham-Cameron Illustration (Graeme Wilson) and Ken Vail Graphic Design

Acknowledgements
Every effort has been made to seek all copyright-holders. If any have been overlooked, then Bright Red Publishing will be delighted to make the necessary arrangements.

The publishers would like to thank the following for the permission to reproduce the following material:
Graphs © BBC Scotland Education (pp 15 & 18); Ordnance Survey © Crown Copyright. All rights reserved. Licence number 100049324 (pages 19, 27, 49, 58); The Scottish Census 2011 logo © Crown Copyright. Reproduced under the terms of the Click-Use Licence (p 38); Photograph © Paul Chesley/Riser/Getty Images (p 43); 1961–2001 Census output © Crown Copyright. Reproduced under the terms of the Click-Use Licence (p 50); Photograph © David Goddard/ Getty Images News/Getty Images (p 52); Photograph © Harvey Nichols (p 56); Photograph © TIE 2009 (p 57); Photograph © Jeff J Mitchell/Getty Images News/Getty Images (p 61); Photograph © Laurence Griffiths/Getty Images Sport/Getty Images (p 61); Photograph © Antony Edwards/The Image Bank/Getty Images (p 62); Photograph © Zack Seckler/The Image Bank/Getty Images (p 64); Image courtesy Jacques Descloitres, NASA/GSFC, MODIS Rapid Response (p 65); Photograph © Margaret Bourke-White/Time & Life Pictures/Getty Images (p 66); Photograph © Daniel Beltra/ The Image Bank/Getty Images (p 68); Photograph © Jeremy Hartley/Panos Pictures (p 69); Photograph © Three Lions/Hulton Archive/Getty Images (p 75); Graph © Dr Jean-Paul Rodrigue, Hofstra University (p 77); Photograph © Glasgow Fort Shopping Park, Junction 10 M8, www. glasgowfort.com (p 78); Photograph © The Royal Bank of Scotland plc (p 79); Photograph © EyesWideOpen/Getty Images News/Getty Images (p 81); Photograph © Gideon Mendel/Corbis (p 89); Photograph © De Agostini Picture Library/De Agostini/Getty Images (p 90); Photograph © Per-Anders Pettersson/Getty Images News/Getty Images (p 90).

Printed and bound in Scotland by Scotprint.

CONTENTS

HIGHER GEOGRAPHY

INTRODUCTION

This book presents the content for the Higher Geography course in a concise and digestible format. It does not attempt to go into the detail that would be found in a full blown textbook. However, it does cover the important examinable concepts and addresses frequently asked questions in a way that is more manageable. If you can learn all the information in this book, and supplement it with the detailed course notes from school, you will be well placed to tackle the Higher Geography exam.

Important words and phrases are highlighted in the text as **key words**. *Internet links* are provided throughout the book. These are not official links but will illustrate a point or provide further interest or understanding of a specific topic. *Don't forget* boxes give key pointers on how to avoid common mistakes, provide examples of what you should do or highlight things that students often struggle with. *Let's think about this* boxes end each spread with a summary of the most important information, ideas on how to research the topic further or tips for the exam.

SYLLABUS

The Higher course consists of three different units – Physical Environments, Human Environments and Environmental Interactions. These are further subdivided into the units shown below.

Physical Environments	Human Environments	Environmental Interactions
Atmosphere	Population	Rural land resources
Hydrosphere	Rural	Rural land degradation
Lithosphere	Industry	River basin management
Biosphere	Urban	Urban change and its management
		European regional inequalities
		Development and health

All aspects of Physical and Human Environments are compulsory – and are referred to as the **core** units. A range of geographical methods and techniques, including interpretation of diagrams and Ordnance Survey map work, is also studied in the core.

The Environmental Interactions consider aspects of both Physical and Human Environments and the interactions that take place between them. You only study two of the six units listed, and these are likely to have already been chosen by your teachers. When revising your chosen interactions, it is important to study the relevant core units at the same time. For example, refer to the relevant parts of the Lithosphere topic when looking at Rural Land Resources and the Urban, Industry and Population topics for Urban Change and its Management.

ASSESSMENT

Unit assessment

Your overall grade in Higher Geography is based solely on your performance in the external exam, which consists of the two papers outlined below. During the year, you have to pass three **unit assessments**, otherwise known as **NABs**. These are marked by your teachers and cover a limited range of topics, so will test you at a lower level than the final exam.

In each of the Physical and Human Environments NABs, you will be asked questions on two of the four relevant topics (for example, Urban and Industry for the Human Environments NAB). Each of these questions is worth 20 marks, and there is an additional 10-mark question on a geographical method or technique.

contd

4

ASSESSMENT contd

For the Environmental Interactions NAB, you will be assessed on one of the topics you have studied. The questions asked will require detailed answers, and part questions can sometimes be worth 20 marks or more.

NABs for all units are out of 50 marks, and you have to score 25 to pass. If you fail, you are entitled to one reassessment (using a different NAB). In exceptional circumstances, and at the discretion of your teacher, you may be allowed a third attempt. As you will probably be sitting NABs in a number of different subjects, it is much better to put some effort into your revision and pass them all first time.

The exam

Paper 1 lasts 1 hour and 30 minutes and covers the eight core topics of Physical and Human Environments. It has three sections:

- **Section A** has four compulsory questions – two each from the four Physical and four Human Environments topics.

- **Section B** will have two Physical questions, and you answer one of them.

- **Section C** will have two Human questions, and you answer one of them.

Although you have some degree of choice in sections B and C, the topics covered vary from year to year, so you cannot ignore any in your revision. The paper is worth 100 marks, so you need to write quickly and concisely to get down as much information as possible. Finally, the paper always contains at least one Ordnance Survey map question, usually in one or more of the following topics: Urban, Industry, Lithosphere or Hydrosphere.

Paper 2 lasts 1 hour and 15 minutes and has one question for each of the Environmental Interactions topics. The two questions you answer are worth 50 marks each. Your answers have to be well developed, as individual parts of each question can be worth up to 26 marks.

 The SQA website has loads more useful information on your exam at www.sqa.org.uk

DON'T FORGET

You have to pass the three unit assessments to be given a course award after the final exam, but NABs do not count towards your final mark.

DON'T FORGET

In sections B and C of paper 1, you are given optional questions to answer – don't attempt to answer them all.

DON'T FORGET

Before 2008, a half-mark, rather than a full mark, was awarded for a valid point. This makes no difference to the way you are assessed – but remember to double the mark scheme when practising exam questions from older papers.

 ## LET'S THINK ABOUT THIS

Writing down enough information to get 100 marks in 75 minutes is a tough task. Timing yourself on past paper questions is an excellent way to prepare and get a feel for what is required in exam conditions.

GLOBAL SCALE: THE HEAT BUDGET

The sun is the main source of energy for the Earth, and the solar energy we receive from it drives the atmospheric system and makes life possible.

Generally, the **incoming solar radiation** (also known as **insolation** or **short-wave solar energy**) is balanced by the **outgoing terrestrial radiation** (also known as **long-wave radiation** or **infra-red radiation**). This balance between input and output is known as the **global heat budget**.

Incoming solar radiation passes through the atmosphere and is transferred to the planet's surface, but:

- some is **reflected** by clouds

- some is **scattered** by gas particles in the air

- some is reflected by the Earth's surface.

This reflection and scattering is called **albedo** and accounts for a loss of around 30 per cent of the incoming solar radiation. In addition:

- some is **absorbed** by clouds (around 3 per cent)

- some is absorbed by dust, water vapour and other gases (around 17 per cent).

So, a total of around 50 per cent of the incoming solar radiation is left to be absorbed by the Earth's surface.

To balance the input, long-wave radiation is **emitted** from the surface back into the atmosphere. Around 6 per cent is lost directly to space while the remaining 94 per cent is absorbed by water vapour, carbon dioxide and other **greenhouse gases** and is then re-radiated. This **greenhouse effect** slows down the loss of radiation from the atmosphere. Without it, global temperatures would be around 30°C cooler.

DON'T FORGET

Reference diagrams you are given in exams may have slightly different figures for albedo, absorption and radiative effects. Do not let this faze you, as the general principles remain the same.

ENERGY RECEIPT

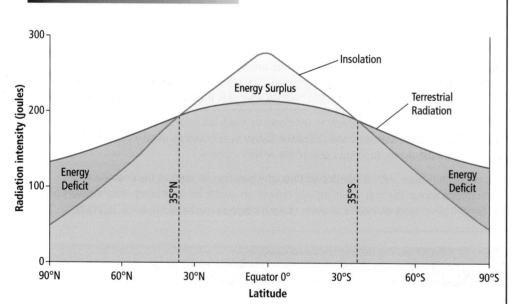

Global energy receipt

Latitudinal variations mean that some parts of the planet receive much more solar radiation than others.

Between around 35°N and 35°S, there is an energy **surplus** because incoming solar radiation is greater than outgoing terrestrial radiation.

contd

ENERGY RECEIPT contd

From 35°N and 35°S towards the poles, there is an energy **deficit** because incoming solar radiation is less than outgoing terrestrial radiation.

The variations in energy surplus and deficit are greatest between the intense heating that takes place in the tropics and the intense cooling that takes place at the poles. The contrast between these two extremes is referred to as the **global temperature gradient** and is caused by the following:

- The **curvature** of the Earth means that the surface of the planet slopes further away from the sun with increasing distance from the Equator. Beams of incoming solar radiation will be concentrated on a smaller area at the Equator, giving higher temperatures than at the poles, where they are more spread out.

- The curvature of the Earth also means that these beams pass through more atmosphere at the poles than at the Equator. They will be affected by more absorption and albedo from clouds, water vapour and gases, so temperatures will be lower.

- Between the tropics of Cancer and Capricorn, the sun is high overhead during the day all year round. Insolation is more focused and temperatures higher than at the poles, where the sun's rays hit the surface at a much lower angle.

- For six months of the year, the poles are covered in darkness, receive little or no incoming solar radiation and have very low temperatures.

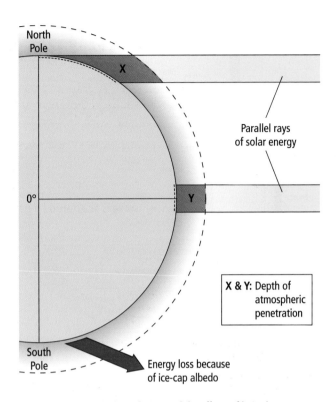

Incoming solar radiation and the effects of latitude

The effects of surface albedo are also important. At the poles, white snow and ice reflect much of the solar radiation back into space. In contrast, at the Equator, there is much more absorption by dense, dark tropical vegetation.

DON'T FORGET

Many students are put off the Atmosphere unit because it appears quite scientific. Don't be – if you follow the basic principles and learn the important terms and concepts, answering questions becomes quite straightforward.

ENERGY TRANSFER

The global differences in energy surplus and deficit would result in a severe energy imbalance, with the tropics becoming gradually warmer and the poles gradually colder, if these were the only processes in operation. To keep the balance, energy is transferred from areas of surplus to those of deficit in two different ways.

Atmospheric Circulation by winds, storms and hurricanes is responsible for the majority of the transfer of energy – around 80 per cent of the total.

The remaining 20 per cent is transferred through **Oceanic Circulation** by ocean currents such as the North Atlantic Drift (or Gulf Stream). Scientists are finding increasing evidence that these currents may be more important in energy transfer than has previously been thought.

More information on the composition of the atmosphere can be found at
http://ssoar.org/university/course/atmSaC.ppt

 LET'S THINK ABOUT THIS

Atmosphere questions are often used to test your interpretation of diagrams. They may look similar to those here or be completely different. Take the time to properly understand the diagram and its key, then quote relevant figures from it in your answer.

GLOBAL SCALE: ENERGY TRANSFER

ATMOSPHERIC CIRCULATION

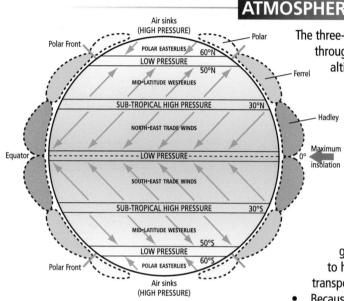

The three-cell model attempts to explain global energy transfer through a series of interconnected cells that move air at high altitudes and have associated surface wind belts:

- Warm air rises at the Equator and travels north and south in the upper atmosphere before cooling and descending at around 30°N and 30°S. This movement forms the **Hadley Cell**.
- At the poles, cold dense polar air sinks and moves towards lower latitudes at the surface, warming as it travels and rising again at the **Polar Front**. This movement forms the **Polar Cell**.
- The **Ferrel Cell** lies between the Hadley and Polar Cells. It is controlled by the other two cells and gains energy from them. Warm air from the tropics is fed to higher latitudes through the Ferrel Cell, and cold air is transported back to the sub-tropics.
- Because the Hadley and Polar Cells are controlled by heat and cold, they are **thermally direct**. The Ferrel Cell is **thermally indirect** because it is controlled by the other two cells.

 Vertical Air Circulation Cell

This model assumes:
- a rotating Earth
- a uniform Earth's surface
- mid-day sun overhead at equinox

Global atmospheric circulation and the three-cell model

As warm air in the Hadley Cell rises at the Equator, an area of low pressure forms at the Earth's surface. Huge **cumulonimbus** clouds form and give rise to heavy **convectional** rainfall. The air spreads north and south and sinks at the **sub-tropical high-pressure areas**. This sinking air is very dry and gives clear skies and little or no precipitation, and it is in these areas that the hot deserts – the Sahara, Kalahari, Atacama and Great Australian Desert – are found.

Rossby waves and the jet stream

Recent evidence has shown that, instead of a Ferrel Cell, the upper atmosphere in the mid-latitudes consists of slowly travelling bands of high and low pressure in horizontal wave-like motions called **Rossby Waves**.

Like rivers, Rossby waves can form meander loops which transfer cold air from the high to low latitudes and warm air in the opposite direction. The waves have associated **jet streams** – narrow bands of very fast-flowing air – that influence mid-latitude anticyclones and depressions.

> **DON'T FORGET**
>
> The three-cell model is a vastly simplified version of reality but is very useful for explaining global energy transfer – make sure you can describe its characteristics in detail.

SURFACE WIND PATTERNS

At the Earth's surface, the effects of the three-cell model are seen in the high-pressure areas at the poles and tropics and the low-pressure areas at the Equator and in the mid-latitudes.

In theory, surface winds should blow directly from high-pressure areas to low-pressure areas. The rotation of the Earth, however, produces the **Coriolis Force**, which deflects winds to the right of their direction of travel in the northern hemisphere and to the left in the southern hemisphere.

Pressure differences, combined with Coriolis, produce distinct global wind patterns:

1. Winds blow from the sub-tropical high-pressure areas (at around 30°N and 30°S) to the Equator, forming the north-east and south-east trade winds.

2. The mid-latitude westerlies flow polewards from the sub-tropical high-pressure areas and have associated series

3. The polar easterlies flow from the polar high-pressure areas to the mid-latitude lows.

of anticyclones and depressions. They are impeded by land masses in the northern hemisphere and are stronger in the south.

The idealised pattern of global surface winds breaks down in reality because of localised conditions such as the monsoon over India and the influence of the Himalayan mountains. It is still useful as an indication of the general pattern.

OCEANIC CIRCULATION

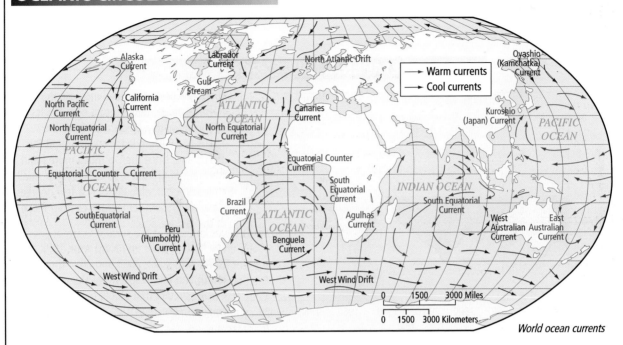

World ocean currents

Energy is also transferred around the planet by ocean currents such as the **North Atlantic Drift** (or **Gulf Stream**), which brings warmer waters to the western coasts of the UK and influences its mild climate. Elsewhere, cold currents off the coasts of Peru and Namibia create dry onshore winds and help to form deserts.

Ocean currents are driven by **thermohaline circulation** – heat and salt determining the density of water. The cold water at the poles is denser because it contains more salt, so it sinks and flows towards the Equator in deep ocean basins. Warmer, less dense surface currents flow polewards from the Equator. The resulting worldwide circulation system is known as the **Atlantic Conveyor**.

The pattern of ocean currents is similar to that of atmospheric circulation, but deflection by land masses results in huge, circular loops of water called **gyres**. These flow clockwise in the northern hemisphere and anticlockwise in the south. The **West Wind Drift** (or **Antarctic Circumpolar Current**) is not impeded by land masses and flows from west to east around the globe.

Other factors responsible for these patterns include:

* The prevailing global wind pattern – surface winds cause friction with the water surface and influence the direction of surface currents

* Deflection by Coriolis, directing currents to the right in the northern hemisphere and the left in the south

* The cold, slow-moving deep-sea currents which have an effect on the surface patterns.

The phenomena known as **El Niño** and **La Niña** – periodic warming and cooling of the tropical Pacific Ocean – also have an influence on global weather patterns, causing extreme events such as heavy rainfall and flooding in normally very dry deserts.

 **DON'T FORGET**

Questions on oceanic circulation, often referring to either the Atlantic or Pacific Oceans, are much less common than ones on atmospheric circulation, but you still need to know the basic characteristics.

 Find more on El Niño from NASA at http://earthobservatory.nasa.gov/Features/ElNino/

LET'S THINK ABOUT THIS

Atmospheric and oceanic circulation may seem complicated at first but can be quite easily explained if you understand why pressure zones are located in certain areas. Learn the major winds and ocean currents and remember basic principles such as the effect of Coriolis and that winds always blow from high to low pressure.

GLOBAL SCALE: CLIMATE CHANGE

GLOBAL TEMPERATURE CHANGE

Recent global temperature change

There has been a general increase in the average global temperature over the last 150 years with a 0·35°C rise from the 1910s to 1940s, a 0·1°C cooling until the 1970s and, since then, a much more rapid warming of around 0·55°C. These are small changes – but, globally, only a few degrees separate present temperatures from those of the last ice age.

The hottest years on record were 1998 and 2005, with 2002, 2003 and 2004 the third, fourth and fifth respectively. Computer models predict that global temperatures could rise between 1·4°C and 5·8°C by 2100.

> **DON'T FORGET**
>
> When analysing global temperature graphs, you need to mention the general trend (whether rising or falling), major and minor changes and prolonged periods above or below the average. Use actual figures from the graph and, very importantly, make sure you understand its scale.

The enhanced greenhouse effect

Greenhouse gases in the atmosphere – including carbon dioxide (CO_2), methane, nitrous oxide, halocarbons and water vapour – keep global temperatures around 30°C warmer than they would otherwise be. They are produced naturally but are added to by human activity. This causes the **enhanced greenhouse effect** which most scientists believe is behind climate change.

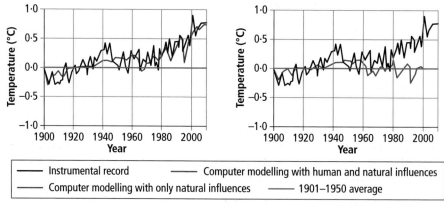

The influence of human activities on global temperature change

Physical causes of temperature change

- **Volcanic eruptions** emit huge amounts of dust into the atmosphere. This is distributed by global winds, forms a blanket shielding the Earth from incoming solar radiation and leads to a decrease in temperature. The effects of eruptions such as Krakatoa (1883) and Pinatubo (1991) can be clearly seen on temperature change graphs.

- **Increased sunspot activity** leads to greater output of solar radiation and increased temperatures, such as those in the warm 1940s. Climate change sceptics have said that this process is responsible for temperature increases over the last 40 years – but, in fact, the sun has cooled slightly since 1970.

- **Milankovitch cycles** (the wobble, roll and stretch theory) are slight variations in the Earth's orbit around the sun. Every 41 000 years, there is a change in the tilt of the Earth's axis, which can bring more sunlight to polar regions. Over a period of 100 000 years, the orbit stretches, increasing and decreasing incoming solar radiation over long periods of time.

- **Changes in oceanic circulation** influence global change although, in the short term, this may be limited to El Niño effects.

contd

GLOBAL TEMPERATURE CHANGE contd

Human causes of temperature change

- **Fossil fuels** (coal, oil and natural gas) were formed over millions of years but are being burned much more quickly in power stations, factories, road vehicles and homes. Since intense burning started with the Industrial Revolution, the amount of CO_2 in the atmosphere has increased around 30 per cent.

- **Deforestation**, especially burning tropical rainforests, adds to the concentration of CO_2 but, more importantly, decreases the amount used by plants in photosynthesis.

- **Methane** produces 21 times as much greenhouse warming as CO_2. It comes from decaying vegetation (especially in rice padi fields), increasingly flatulent cattle and waste in landfill sites. **Nitrous oxide** comes mainly from agricultural fertilisers and pesticides. Sources of these gases are increasing rapidly in line with global population increases.

- **Halocarbons** (e.g. **chlorofluorocarbons** or **CFCs**) are up to 13 000 times more potent than CO_2. They come from solvents and fridges but were banned internationally in 1989. They can remain in the atmosphere for up to 400 years.

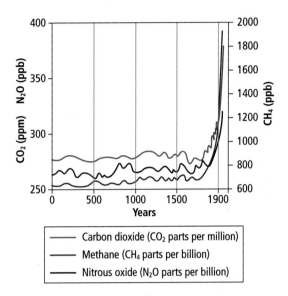

Increasing concentrations of important greenhouse gases

CONSEQUENCES OF CLIMATE CHANGE

There is great debate on the effects of global warming. Possibilities include:

- Sea-level rises of between 9 cm and 69 cm by 2080, leading to coastal flooding and some island nations (e.g. the Maldives) becoming uninhabitable

- An increase in extreme weather events such as European heatwaves, Australian forest fires and severe hurricanes in the USA

- The disappearance of the polar ice cap

- The extinction of up to a third of all species by 2100

- Tropical diseases (e.g. malaria) moving into temperate latitudes

- Climate wars fought over precious food and water

- Bleaching of coral reefs

- Changing agricultural patterns as, for instance, potatoes give way to maize in Scotland.

Long-term effects may include: much larger sea-level rises and the weakening of the Gulf Stream if the Antarctic and Greenland ice sheets melt; an increased release of methane from melting Arctic permafrost accelerating the global rise in temperature; a loss of locally important economic activities such as skiing in the Alps.

Some potential impacts, such as warmer Scottish summers, are positive, but many are negative and will have the greatest effect on developing countries and the world's poorest people.

DON'T FORGET

Examiners love up-to-date information, especially when you can relate it to graphs and tables. Learn the dates of some volcanic eruptions and periods of sunspot activity to be really impressive in your answers.

 There is a wealth of up-to-date information on climate change at http://news.bbc.co.uk/1/hi/sci/tech/portal/climate_change/default.stm

LET'S THINK ABOUT THIS

The causes and effects of global climate change are very varied. Make sure you know the physical and human sources of greenhouse gases and can relate them to specific periods shown on graphs. Don't just concentrate on CO_2 – other gases may exist in lesser quantities but, relatively speaking, be far more dangerous.

REGIONAL SCALE: WEST AFRICA

CLIMATE AND VEGETATION

West Africa has a tropical climate with temperatures greater than 23°C all year round. With increasing distance from the Equator, there is a greater range of temperature, decreasing rainfall and increasingly distinct wet and dry seasons.

The climate between the Equator and 10°N and S is **equatorial**. Temperature and rainfall are both high, with little variation throughout the year. This gives rise to natural rainforest vegetation.

Further north and south, the climate is **tropical continental**, often called **savanna**. Temperatures are high, rainfall is lower and there is a distinct pattern of wet and dry seasons. Vegetation is savanna grassland or scrub and decreases with increasing distance from the Equator.

Beyond the savanna lies the **hot desert** climate (the Sahara in northern Africa). Daytime temperatures are hot with a large **diurnal** range, and rainfall is less than 250 mm a year. Vegetation is minimal, and farming is almost impossible.

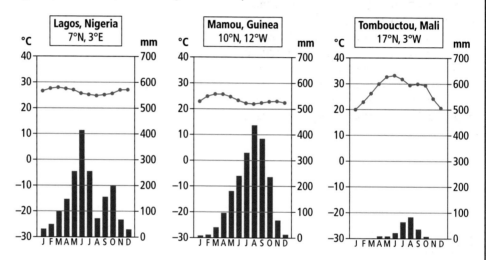

Climate graphs showing increasing distance from the Equator in West Africa

THE INFLUENCE OF AIR MASSES AND THE ITCZ

Large masses of air settle over parts of the Earth's surface and take on the characteristics of the areas where they form. When these **air masses** move, in the form of surface winds, they carry the characteristics of their **source areas**. There are two main air masses affecting the weather and climate of West Africa.

Tropical continental

The **cT** air mass is formed over the Sahara desert, and its associated winds are the **north-east trades**, known locally as the **Harmattan**. These bring extremely hot temperatures, a very low relative humidity between 10 and 17 per cent, no rainfall and poor visibility because of the large amount of dust they carry.

Tropical maritime

Formed over the ocean south of the Equator, the **south-east trade winds** from the **mT** air mass are deflected by Coriolis as they pass over the Equator to become the **south-west monsoon**. These winds are hot, having formed in a tropical area, but humidity is high (from 65 to 82 per cent). Associated rainfall varies from showers to intense thunderstorms.

> **DON'T FORGET**
>
> Questions on the ITCZ and West Africa are quite common, and you can pick up easy marks for learning the characteristics of air masses. Remember that all of West Africa is hot, even though cT winds tend to be hotter than mT ones.

contd

THE INFLUENCE OF AIR MASSES AND THE ITCZ contd

The inter-tropical convergence zone

Areas where different winds meet are called **zones of convergence** (a **zone of divergence** is where winds blow away from each other). The zone where the north-east trade winds and the south-west monsoon meet over West Africa is called the **inter-tropical convergence zone** (or **ITCZ**). The ITCZ is a zone of **low pressure** where air rises. South of the ITCZ, a wide band of different types of rainfall forms over several hundred kilometres.

The ITCZ is not stationary but travels north and south of the Equator following the overhead sun and the **thermal equator** – the area of most intense heating of the Earth's surface by the sun. In July (the northern hemisphere 'summer), the thermal equator is at its most northerly extent, around 20°N, because the tilt of the Earth means that the Sahara is receiving maximum incoming solar radiation. It then migrates south towards the Tropic of Capricorn until the peak of the southern hemisphere summer in January. However, part of the ITCZ remains at around 8°N over West Africa because of the differential heating of the oceans and the land surface.

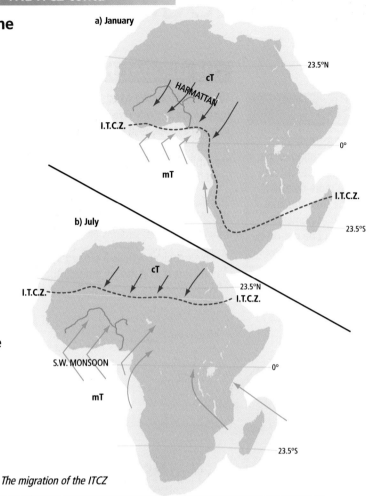

The migration of the ITCZ

WEST AFRICAN RAINFALL AND THE ITCZ

The position of the ITCZ determines the rainfall pattern over West Africa. In January, it is located near the southern coast of West Africa, and places such as Lagos in Nigeria receive light rainfall. Areas north of the ITCZ are affected by the hot, dry winds of the cT air mass and experience their dry season.

As the ITCZ migrates north, large cumulonimbus clouds follow to the south and bring thunderstorms over Lagos and further inland. South of the thunderstorms, rainfall will decrease once more, leading to a dip in monthly totals. As the ITCZ reaches its northernmost point in July, all areas south of around 20°N have their wet season.

The ITCZ then migrates south, and northern areas become dry once more. Southern coastal areas in West Africa will experience a returning band of thunderstorms and high rainfall, leading to a second rainfall peak. These areas, which are always south of the ITCZ, do not have a dry season.

 **DON'T FORGET**

You will often be asked to interpret West African rainfall and temperature data, possibly in the form of a climate graph, so make sure you can relate your knowledge to it.

 Move the slider and watch the progression of the ITCZ at http://daphne.palomar. edu/pdeen/Animations/23_WeatherPat.swf

LET'S THINK ABOUT THIS

Rainfall patterns across West Africa depend on air masses and the position of the ITCZ. In general, the further north the more rainfall is concentrated in the summer months (August being the peak month), the smaller the total mean annual rainfall and the more variable that rainfall is. This has had a big impact on drought in the Sahel region on the southern edges of the Sahara desert.

HYDROLOGICAL SYSTEMS

THE GLOBAL HYDROLOGICAL CYCLE

DON'T FORGET

There are lots of different processes operating in the hydrological cycle, but it is basically a more detailed water cycle than the one you learned at primary school – don't be put off by the more scientific terms.

The mechanisms by which water is transferred between land and sea are shown by the **global hydrological cycle**. Water is found on the Earth's surface in oceans, seas, lakes, rivers and streams as well as in snow, glaciers and ice caps. Heat from the sun causes **evaporation** from these sources, and water enters the atmosphere as **water vapour**. It is added to by **transpiration** from plants. Water vapour is moved around the planet by the wind, cools and condenses to form clouds and is returned to the surface as **precipitation**.

This water makes its way back to the oceans by entering the soil as **infiltration**, moving through the soil as **throughflow** and **percolating** downwards to become **groundwater**. Some is also returned to rivers, streams and lakes as **surface runoff**.

The hydrological cycle is a **closed system** because the amount of water it contains is constant – none is lost as it moves through the system.

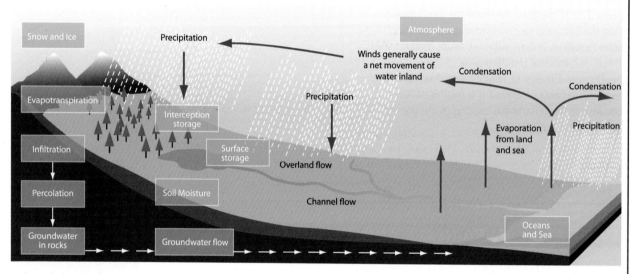

The global hydrological cycle

THE DRAINAGE BASIN SYSTEM

All rivers have **tributary** streams. The **catchment** area drained by a river and all its tributaries is called the **drainage basin**. The boundary of the drainage basin is the edge of higher ground that separates it from its neighbouring basin and is called the **watershed**. Some drainage basins, such as the Amazon and the Mississippi, are huge but many are quite small.

The drainage basin is an open system with **inputs**, **processes** (storage and transport) and **outputs**.

Inputs

Precipitation is the main input, but it is important to know the type (snow, sleet, hail and rain), intensity, frequency and duration.

Processes

Water storage processes include: **interception** where leaves temporarily hold precipitation before it falls to the ground; **surface storage** in everything from puddles to lakes; **soil moisture storage** within the ground; and **groundwater storage** in the rocks underneath the soil. Most water eventually ends up in streams and rivers as **channel storage**.

contd

THE DRAINAGE BASIN SYSTEM contd

When water is released from storage, it is transported by a number of different processes:

- **Throughflow** as water falls from leaves to the ground, and **stemflow** as it drips off branches and tree trunks

- **Infiltration** into the soil, and **soil throughflow** downhill under the influence of gravity

- **Percolation** deeper into the soil and underlying rock to become **groundwater flow**

- **Runoff** from the land as **overland flow** (surface runoff) or **streamflow** in the river channel itself.

Some precipitation will also enter the river system directly as **channel precipitation**.

Outputs

Some water is lost through **evapotranspiration** – the combined total of all the water **evaporated** from soil and water surfaces and lost from plant leaves through **transpiration**.

The rest of the water makes its way to the sea and leaves the system as **drainage basin output**.

DON'T FORGET

You have to be able to annotate the various parts of a hydrograph and also analyse what they tell you – more on this can be found later in the chapter.

STORM HYDROGRAPHS

To compare rivers, we need to look at their **discharge** – the amount of water in the river passing a certain point every second. Discharge is measured in **cumecs** (cubic metres per second) and is calculated by multiplying the speed of the river by its cross-sectional area.

At various points in a drainage basin, the discharge will be measured at a **gauging station**. It can then be compared with precipitation by drawing a **storm** (or **flood**) **hydrograph**. Features shown on a hydrograph include the following:

- **Precipitation** – normally related to a specific event such as a storm and shown against a **time** scale on the x-axis

- The normal level of the river, known as the **baseflow**

- The **discharge** of the river, also shown against time

- The **peak flow** discharge for the storm period

- The gap between peak rainfall and peak flow, known as the **basin lag time**. This reflects direct channel precipitation plus the overland flow and throughflow, which takes longer to reach the river

- The **rising limb**, which shows how quickly water reaches the river channel

- The **recessional** or **falling limb**, which shows the decrease from peak flow

- The **storm flow** – the additional discharge produced by the storm event

- The **bankfull discharge level** above which the river will be in flood.

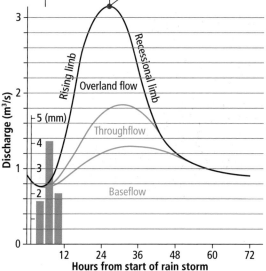

An example of a flood hydrograph

DON'T FORGET

Rainfall on the hydrograph will be shown as a bar chart with the discharge as a line graph. There will be two separate scales on the y-axis for the rainfall and discharge, so be sure to distinguish between them.

 Practise drawing hydrographs in an Excel spreadsheet using www.ltscotland.org. uk/Images/Hydrographs_tcm4-302099.doc

LET'S THINK ABOUT THIS

There are lots of new terms to be learned in the Hydrosphere unit, but many of them are repeated in the hydrological cycle, the drainage basin system and in storm hydrograph studies, so you just need to learn how to apply them in the three different situations.

THE RIVER PROFILE AND THE FORMATION OF RIVER FEATURES

A river has three stages of development as it flows from **source** to **mouth**:

1. The **youthful stage** or **upper course** with a fast-flowing stream and mainly **vertical erosion**

2. The **mature stage** or **middle course** where river width increases and erosion is more **lateral**

3. The gentle gradient of the **old stage** or **lower course** where **deposition** takes over from erosion as the major process.

RIVER PROCESSES

Along a river's profile, processes of erosion, **transportation** and deposition vary in their importance and give rise to distinctive hydrological formations.

Erosion

Erosion deepens, widens and lengthens the river channel by different methods.

- **Abrasion** or **corrasion** is where the **load** of the river – the sand, pebbles and stones it carries – wears away the bedrock and banks, particularly when the upper course floods.

- **Corrosion** is the erosion of the sides and bedrock of the river by the action of chemicals, such as carbon dioxide, dissolved in the flowing water.

- **Hydraulic action** results from the pressure of flowing water dragging particles from the sides of the river and is mainly responsible for lateral erosion.

One further erosion process, **attrition**, takes place when the particles making up the load rub against each other and become smaller and more rounded.

Transportation

The load of the river comes from the surrounding hillsides, as well as the channel itself, and is transported in four different ways:

1. **Saltation** – relatively small particles on the river bed are dislodged by other particles, travel for a short distance in a bouncing motion, then settle again.

2. **Solution** – the products of chemical erosion are dissolved and carried in the water.

3. **Suspension** – clay and silt, the finest particles, are carried along suspended in the water.

4. **Traction** – the largest stones and pebbles are rolled and dragged along the bed by the flowing water.

Due to erosion, the size of the load decreases downstream and, ultimately, an average of 75 per cent is carried in suspension. Larger particles may only be carried in high flow conditions.

Deposition

When the velocity of a river decreases, it can no longer transport its load, and deposition takes place. This decrease may occur when a river enters a loch or the sea and forms a delta, when a river floods its banks, on the inside of a meander or where the gradient suddenly decreases (e.g. immediately below a waterfall).

DON'T FORGET

Knowing your processes is only half the story – you need to relate erosion, transportation and deposition to the formation of a variety of river features.

RIVER LANDFORMS

The upper course

The river has a steep-sided, **v-shaped valley** with a steep gradient and the channel occupying all or most of the valley floor. The river flows around **interlocking spurs** formed by sloping valley sides. Erosion is mainly vertical, and the bedload consists of large particles.

Waterfalls tend to form in the upper course where a band of resistant rock crosses the river channel and the softer rocks downstream are eroded more rapidly. Erosion is at its greatest under the waterfall where a **plunge pool** forms and, as the softer rock is eroded, it undercuts the harder rock above, causing it to collapse. This process repeats over time and the river retreats upstream, forming a **gorge**.

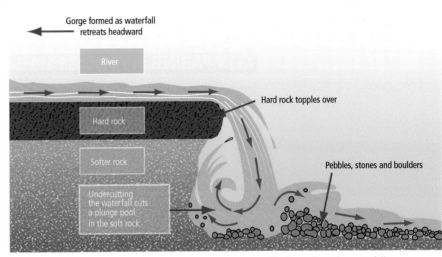

The formation of a waterfall

The middle course

The river channel is wider and deeper with a well-developed flood plain and gentler valley sides. Vertical erosion still takes place but lateral erosion is more important, causing the river to bend and spread across its flood plain. The size of the bedload decreases and suspension becomes a more important form of transportation.

Water swings from side to side in the river channel around deep stretches of slow-moving water (**pools**) with faster shallower sections (**riffles**) in between. Curves known as **meanders** develop as the river moves horizontally. Erosion undercuts river cliffs on the outside of the bend, and deposition forms slip-off slopes on the inside. Meander formation is assisted by **helicoidal** flow, a corkscrew-like movement of water.

Cross-section of a meander

Meanders migrate downstream and change constantly. Large meander loops may be cut off at the neck during times of flood, creating **oxbow** lakes.

The lower course

The channel is at its broadest and deepest. All the bedload is transported in either suspension or solution and deposition dominates.

The river meanders across a wide flood plain, covered in **alluvium** from flooding. Natural ridges or **levées** build up on the edge of the channel. Deposition in the river causes **braiding** and huge dumps of material when the river enters a loch or the sea form **deltas** and **estuaries**.

DON'T FORGET

Rejuvenation gives a river a fresh lease of life when, for instance, sea level is lowered. River terraces, knickpoints and incised meanders are all features of this process.

An interactive Hydrosphere PowerPoint can be found at http://lanarkgrammargeography.pbwiki.com/Hydrosphere

LET'S THINK ABOUT THIS

A lot of the work on river features and processes follows on from what you did at Intermediate 2 and Standard Grade. Much more detail is needed, however, about the formation of the following features: waterfalls; pools, riffles and meanders; oxbow lakes; flood plains and levées; estuaries and deltas.

DRAINAGE BASIN PATTERNS

COMPARING HYDROGRAPHS

Storm hydrographs have different shapes depending on the influence of both natural factors and human activity.

Size and shape

Large drainage basins have greater runoff because they receive more precipitation. Lag time is longer because it takes more time for all the water to reach the main channel. They also tend to have a higher **drainage density** – more streams per square kilometre – and, therefore, more opportunities for precipitation to enter the main channel quickly.

Precipitation in circular basins reaches the main channel much more quickly than in elongated ones, so peak flow will be higher. Gradient is important, as steep slopes will shed water into the channel much more quickly than gentle ones, increasing peak flow.

Soil type

Thick soils have greater infiltration rates than thin soils, increasing lag time. **Impermeable** soils such as clay will increase surface runoff, meaning that water enters the main channel faster than in sand-based **permeable** soils.

Underlying rock type

Apart from their effect on soil formation, the underlying rocks have a big influence on surface and groundwater flow. Impermeable rocks will increase soil throughflow and surface runoff by preventing infiltration. Permeable rocks will let water through but, if it travels along the joints and bedding planes in limestone, it may take some time to reach the channel.

Surface drainage is also low on permeable rocks, giving less opportunity for precipitation and surface runoff to directly enter the main river system.

Human impacts

Increased urbanisation means that large areas of previously rural or agricultural land have been covered with concrete and tarmac, which act in the same way as impermeable rock, increasing surface runoff. Sewers and drains may also quicken the rate at which water enters river channels, especially in times of flood. Building on arable land, and the removal of crops that were important in intercepting rainfall, have led to higher peak flows.

Forests intercept large amounts of precipitation in their canopies and also transpire and take water up in their roots. **Afforestation** delays rapid runoff, while deforested basins are more prone to flooding.

DON'T FORGET

You may be asked to describe and explain different storm hydrographs. Use figures on both precipitation and discharge from the graphs you are given and relate them to the full range of influences described here.

DON'T FORGET

Questions on the effects of shape and steepness of river basins on storm hydrographs have never been asked in the external exam but may come up in your NAB.

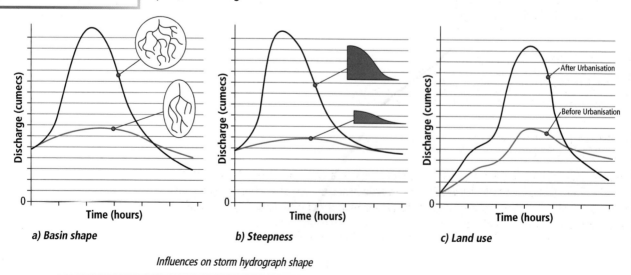

a) Basin shape

b) Steepness

c) Land use

Influences on storm hydrograph shape

RIVER PATTERNS ON ORDNANCE SURVEY MAPS

As well as the common features associated with the upper, middle and lower courses, a description of a river from an Ordnance Survey map should include:

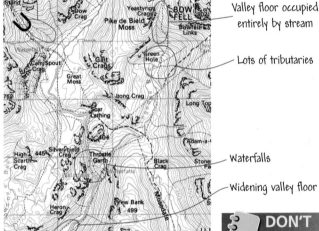

Distance between contours shows wide, flat flood plain

Embankment/levée

Large meanders

Ox-bow lake   *Confluence point*   *Braiding*

Features of a river in its upper course

- whether it is in the upper, middle or lower course (youthful, mature or old)

- a general description of whether it is straight or meandering

- the direction in which it is flowing

- the number and direction of flow of its tributaries

- the changing width of the valley floor

- information on the depth and steepness of the valley sides

- an indication of height to show how steep the gradient of the river is and to give an idea of what course it is in

- appropriate grid references to identify the location of important features.

Unless specifically asked for, there should be no mention made of human features.

In the lower course, additional features may become obvious such as marsh land forming on wide flood plains and the influence of the tide if the river runs into the sea. These features can be easily identified using the OS key.

Contour pattern indicates a V-shaped valley

Valley floor occupied entirely by stream

Lots of tributaries

Waterfalls

Widening valley floor

Features of a river in its lower course

DON'T FORGET

You are often asked to identify river features from maps and then describe their formation – remember to use the diagrams you have learned for this, even if you are not specifically asked for them.

An example of an Ordnance Survey Hydrosphere question can be found here:
http://www.bbc.co.uk/scotland/education/bitesize/higher/img/geography/physical/hydrosphere/geo_hydrosphere_osmap.pdf

 ## LET'S THINK ABOUT THIS

Ordnance Survey map questions on rivers are very common, as are ones in which you have to describe river features. Questions involving the hydrological cycle and storm hydrographs also occur every three or four years. You have studied rivers before, but remember that you need much more detail for full marks at Higher.

WEATHERING AND UPLAND LIMESTONE

Weathering is the breakdown and decomposition of rocks **in situ**. There are two main types.

PHYSICAL WEATHERING

Rocks are broken down into smaller fragments with no change in their chemistry:

- **Frost-shattering (freeze–thaw action)** occurs in cold areas. Water enters cracks and joints in rocks, expanding as it freezes at night and contracting when it melts in the day. This continual stress eventually causes the rock to fracture and break.

- **Exfoliation (onion-skin weathering)** takes place in warmer climates where a large diurnal temperature range causes expansion and contraction of the rock surface until it starts to peel off in layers.

Biological weathering occurs when plant roots growing in cracks in rocks cause the rock structure to break.

CHEMICAL WEATHERING

Rocks are decomposed by chemical action, especially in wet and warm climates. Processes include **hydrolysis, hydration** and **oxidation** – this last turning the iron found in many rocks to rust.

Solution is the main form of chemical weathering and greatly affects limestone, an **alkaline** rock susceptible to attack by rainwater, a dilute **carbonic acid**. Chemical reactions dissolve the rock (a process known as **carbonation**), which is then carried in solution by the water.

UPLAND LIMESTONE

Carboniferous limestone was formed over 300 million years ago when warm shallow seas covered the UK. The rock is around 80 per cent **calcium carbonate (CaCO$_3$)**, made up of the compressed remains of the skeletons of dead sea creatures that accumulated on the sea floor.

Limestone is a hard, resistant rock but is fractured and divided into blocks based upon horizontal **bedding planes** and vertical **joints**. These lines of weakness make limestone **permeable** by

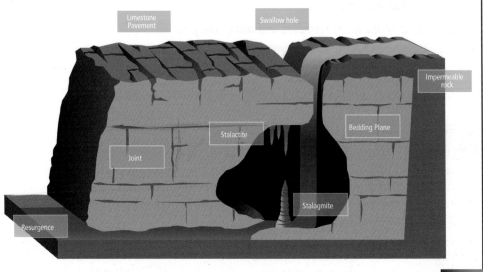

Features of a limestone area

contd

UPLAND LIMESTONE contd

allowing water to pass through. This permeability means that there is very little **surface drainage** in limestone areas, but complex cave systems containing water exist underground – this is known as **intermittent drainage**.

Weathering of surface limestone, joints and bedding planes forms distinctive limestone scenery, known as **karst**, as found in the Yorkshire Dales National Park.

Surface features in limestone landscapes

In upland areas, much of the limestone has been affected by glaciation. This has stripped the vegetation and overlying soil, leaving the limestone open to weathering. On valley sides, limestone **scars** have been formed. When horizontal bedding planes have been exposed on the tops of hills, **limestone pavement** is the characteristic feature. Physical weathering and solution have widened the surface joints to form the **grykes** that separate blocks of limestone, known as **clints**.

Enlarged joints may have filled with glacial deposits to form hollows called **dolines** or **shake holes**. If rivers or streams flow off nearby **impermeable cap rock** and wash away the debris in these shake holes, the water source may then disappear underground, eroding the joint further to form a **swallow hole** or **pothole**. These subterranean streams can flow for long distances underground using joints and bedding planes until they find impermeable rock and emerge at the surface as a **spring** or **resurgence**.

There are two other main features known as **glaciokarst**. These were formed when the ground was frozen and, therefore, acted in the same way as an impermeable layer of rock. Huge quantities of meltwater eroded the limestone into river valleys – but, as the glaciers melted, these became **dry valleys** and **gorges** as water flowed underground once more.

DON'T FORGET

The most important thing about upland limestone is the chemical reaction that takes place between the rock and slightly acidic precipitation – it is this reaction that is responsible for nearly all the features found in a limestone area.

DON'T FORGET

Describing the processes that form surface features leads naturally on to describing the formation of features underground.

UNDERGROUND FEATURES IN LIMESTONE LANDSCAPES

Subterranean streams take water deep under the surface in limestone areas, enlarging joints and bedding planes to form caves. As water enters these caves, the different atmosphere causes it to lose CO_2 and a white mineral known as **calcite** is **precipitated** out. This forms **dripstone** deposits of:

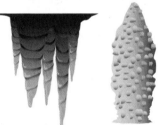

Stalactites, stalagmites and pillars

- **flowstone** – sheet-like deposits where water flows down walls or across the floor of caves

- **stalactites** – 'icicle'-like calcite deposits left behind as water drips from the roof of the cave

- **stalagmites** – thicker, rounded deposits formed where water hits the cave floor

- **pillars** or **columns** where stalactites and stalagmites join.

Sections of rock underground, weakened by the action of water, may collapse and form huge **underground caverns**. Sometimes, roof collapse continues right to the surface – another way in which gorges are formed.

DON'T FORGET

Stalactites hang down from the roof, and stalagmites grow up from the floor. Think of a rhyme to help you remember – or ask your teacher!

 The BBC website on limestone in the Yorkshire Dales is pretty unbeatable: http://www.bbc.co.uk/scotland/education/int/geog/limestone/index.shtml

LET'S THINK ABOUT THIS

When describing the formation of features in a limestone landscape, go right back in time and explain how layers of dead sea creatures built up in warm, shallow seas. Everything else stems from the rock structure that developed because of this.

GLACIATED UPLANDS

The UK has been affected by successive ice ages for 2·4 million years, with the last glacial ice disappearing around 8000 years ago. The processes of glaciation have left distinctive formations of erosion and deposition across upland areas, including the Scottish Highlands and the Lake District.

GLACIAL EROSION

Processes

Glaciers erode very effectively using two main processes:

1. **Plucking** – glacial ice freezes on to bedrock and, when it moves, pulls away pieces of rock, which are then added to the sides and base of the glacier

2. **Abrasion** – rocks incorporated into the glacier scrape away at valley floors and sides, leaving grooves and smooth surfaces behind.

Freeze–thaw weathering is also important, as it fractures rock and creates debris which falls on to glaciers and increases their erosional force.

Landforms

Glacial erosion forms distinctive features in upland areas.

Corries (**coires, cirques** or **cwms**) are deep basins cut into mountainsides with steep **backwalls** and a **rock lip** that may trap water, forming a **corrie lochan** or **tarn**. Most are north-facing and start as a patch of snow building up in a small hollow. Over time, a glacier begins to form and, under gravity, flows downhill through **rotational sliding**. Plucking steepens the backwall, and abrasion erodes and deepens the bottom of the basin. Erosion at the front edge of the corrie is less powerful, and the rock lip develops.

When two corries form back to back, an **arête** (or **knife-edged ridge**) develops as the backwalls meet. Freeze–thaw further steepens this ridge. If three or more corries develop back to back, the same processes form a **pyramidal peak**.

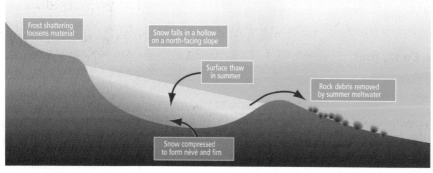

Formation of a corrie

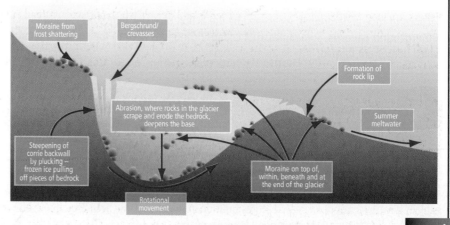

contd

GLACIAL EROSION contd

As glaciers leave the corries, they flow down former v-shaped river valleys widening, deepening and straightening them into steep-sided **glacial troughs** (also known as **u-shaped valleys**). The interlocking spurs of the v-shaped valley may be cut off as **truncated spurs** with smaller **hanging valleys** in between. Water flowing in hanging valleys often exits by a waterfall, while glacial troughs contain long, narrow and deep **ribbon lochs** and **misfit streams** – rivers that appear too small for the u-shaped valley they run in.

Other features of glacial erosion include **roches moutonnées** and **crags-and-tails**.

DON'T FORGET

If asked to describe the formation of one glacial feature, make sure you use a corrie. Formation diagrams are simple to draw, and you can note down lots of detail.

GLACIAL DEPOSITION

Processes

Glacial **transportation** can take eroded material a very long way. Evidence for this exists in the form of **erratics** – large rocks found considerable distances from their source areas.

A mix of rocks, clays, sands and silts is added to the moving glacier by weathering and erosion processes. When the glacier melts, all this **englacial** material is deposited as one of two types of glacial **drift**:

1. **Unsorted** and **unstratified** material deposited directly by the ice is called **till**.

2. **Sorted** and **stratified** material deposited by the **meltwater** flowing out of the glacier forms **fluvio-glacial deposits**.

Till deposits

Smooth, elongated mounds of till are known as **drumlins**. They have a steep, blunt **stoss** end which faces the direction of ice movement and a much more streamlined **lee** end downstream. They usually occur in large numbers as **swarms**.

Various ridge-like deposits of till are known as **moraine**.

- **Terminal** moraine is found at the maximum extent of the glacier and forms across a valley.

- **Lateral** moraines develop along the sides of a glacier and are often composed of material falling on top of the glacier as a result of freeze–thaw.

- When two glaciers meet, the lateral moraines combine to form a **medial** moraine.

- **Recessional** moraines develop when the **retreat** of a glacier is halted. They often form parallel to the terminal moraine.

DON'T FORGET

Glaciers can move forwards or backwards as the climate warms up and cools down. **Advancing** glaciers move forwards and **retreating** glaciers move back.

Fluvio-glacial deposits

Large meltwater streams form **outwash plains** beyond the terminal moraine. Heavier materials such as gravels are deposited first, with fine sands further downstream. Water-filled depressions known as **kettle-hole lochs** can develop if large, buried pieces of ice melt as the glacier retreats. Not all **kettle holes** fill with water.

Eskers are long, winding ridges of stratified sands and gravels which originally formed in tunnels underneath the glaciers in meltwater streams. **Kames** are small hillocks of fluvio-glacial material which built up as small deltas along the ice margins or accumulated in crevasses. If such material accumulated in meltwater streams at the sides of the glaciers, **kame terraces** may be formed.

Find out more about glaciers in general from the USA's National Snow and Ice Data Center at http://www.nsidc.org/glaciers/ and more specific details on landforms here http://www.fettes.com/Cairngorms/glacial%20landscapes.htm

LET'S THINK ABOUT THIS

There are a large number of features of glacial landscapes that you could be questioned on, and you should try to revise them all – if you can't, make sure you know at least two erosional landforms and two depositional landforms in detail.

COASTLINES OF EROSION AND DEPOSITION

Waves play the most critical role in shaping our coastline – eroding, transporting and depositing material. They are produced by **friction** as wind blows over the surface of the sea. How strong they are depends on:

- the strength of the wind

- the duration of the wind

- the distance over which the wind blows, known as the **fetch**.

When waves crash on the shore, water and the material it contains is thrown up the beach as **swash**, while that returning to the sea is called **backwash**. **Constructive** waves have a strong swash and produce a gently sloping beach, while **destructive** waves with strong backwash remove material and form a steep beach.

COASTAL EROSION

Processes

- As waves hit cliff faces, **hydraulic action** traps air in cracks, compressing it and putting stress on the rock, eventually causing it to fracture and break.

- **Abrasion** (or **corrasion**) occurs when stones, sand and other materials are thrown against a cliff face by waves.

- Boulders and pebbles bump and grind away at each other and become smaller particles through **attrition**.

- Weathering by the chemicals found in seawater is called **corrosion**.

As well as the erosion caused by wave action, **sub-aerial** processes of mass movement and weathering such as freeze–thaw will be taking place above the high tide mark.

DON'T FORGET

When describing coastal erosion processes, use lots of detail – describe hydraulic action and attrition, don't just say wave action – and go back to the beginning. So, when describing the formation of a stump, start with headland development.

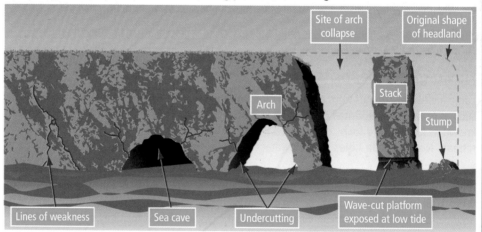

Cracks, caves, arches, stacks and stumps

Landforms

Alternate bands of hard and soft rock are found all along the coastline of the British Isles. The resistant bands of hard rock are eroded much more slowly than the soft rock and are left protruding as a succession of **headlands** with indented **bays** between them.

Wave action attacks the foot of a cliff between the high and low tide marks and erodes a **wave-cut notch**. Eventually, the overhanging cliff face will collapse. The process then repeats, and the cliff becomes steeper and higher with an increasingly large **wave-cut platform** at its base.

contd

ures to develop on cliffs and headlands:

es.

he roof
ole.

cut its way

- The roof of the arch may become unstable as it is attacked by wave action and sub-aerial processes. If it collapses, a **stack** is left standing alone at sea. Erosion of the foot of the stack will cause it to fall, leaving a **stump**.

 falls, another may be in the process of

_____

rosional
and

ut
t gradient,
t. Over
in this way

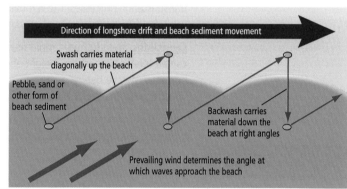

Direction of longshore drift and beach sediment movement

Swash carries material diagonally up the beach

Pebble, sand or other form of beach sediment

Backwash carries material down the beach at right angles

Prevailing wind determines the angle at which waves approach the beach

Longshore drift

hen the energy of the waves decreases. The most
e often develop in sheltered bays where wave

lines. If the coastline changes direction, wave
l may take place in open water. Three features are

n water and can extend for several kilometres beyond
m a **lagoon** or may become **salt marsh**. Spits are in
ably over short periods of time (decades).

mainland further down the coast. Again, lagoons
andbars.

sland out at sea, they are known as **tombolos**.
w tide.

DON'T FORGET

Quoting famous case study examples of spits (Spurn Head), stumps (Old Harry) and other formations could get you extra marks.

as ice melts at the end of an ice age, while
d relative sea level falls as the weight of the ice is
atures including **raised beaches, raised wave-cut platforms, raised deltas, raised mudflats, rias** and **fiords**.

Another interactive BBC site, focusing on the north-east Scottish coastline, is
http://www.bbc.co.uk/scotland/education/geog/coastline/enhanced/

 LET'S THINK ABOUT THIS

As with all of the Lithosphere landscapes, being able to identify the features may mean little without detailed knowledge of the processes behind their formation. Learn important terms well and know the difference between deposition and erosion.

LA'GOON

SPIT

BEACH

LONGSHORE DRIFT

MASS MOVEMENTS

The processes which move large amounts of soil, stones and rock (known as **regolith**) downhill under the force of gravity are known as mass movements. They are influenced by:

- the type of rock and its structure – whether it is porous or impermeable, weak or strong, jointed or sloping

- how much water is present and whether or not the debris is saturated

- the amount of surface vegetation

- the angle of slope – steep slopes encourage faster movements.

Mass movements are important in landscape formation and often work with other erosion and transportation processes. The removal of material produced by a rockfall, for instance, may be carried out by a river. Types of mass movement include:

Landslides and slumps – rapid movements of regolith on slopes when the underlying rock can no longer support the weight above. They occur where saturated bedding planes, lubricated by heavy rain, lie between contrasting layers of rock. Slumps occur in weaker rocks when **rotational movement** is involved.

Rockfalls – very rapid movements on very steep slopes or cliffs when freeze–thaw or exfoliation processes are at work. At the foot of the cliff, debris will accumulate as a **scree** or **talus** slope.

DON'T FORGET

Questions solely about mass movements are rare (although not unknown), but knowing how they are caused may prove useful in answering other Lithosphere questions.

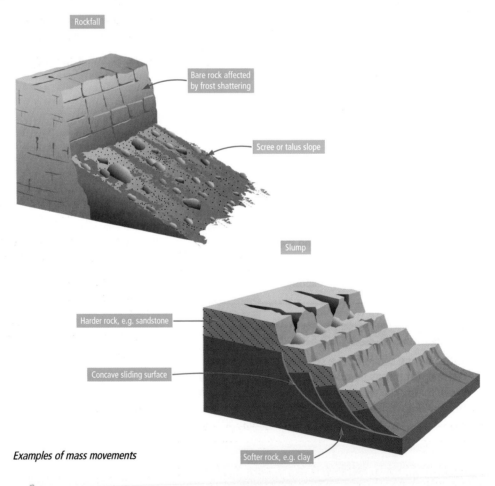

Rockfall

Bare rock affected by frost shattering

Scree or talus slope

Slump

Harder rock, e.g. sandstone

Concave sliding surface

Examples of mass movements

Softer rock, e.g. clay

 Details of the disaster caused by a mass movement at Aberfan in Wales can be found at http://www.nuffield.ox.ac.uk/politics/aberfan/home.htm

LITHOSPHERE ON ORDNANCE SURVEY MAPS

A stream, previously flowing over impermeable cap rock, disappears down a swallow hole

Shake holes may be quite obvious

Limestone pavement on flat ground

Potholes

Resurgence where water emerges from underground streams

Steep sides with no river = dry valley

Scars on valley sides

Names may indicate areas used as case studies

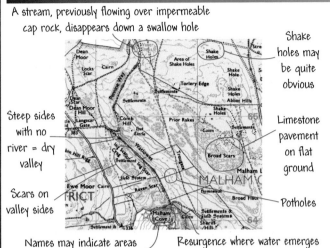

Upland limestone

- Limestone pavement and scars look quite similar on maps so need to be distinguished using contours.
- There is no surface water except on intrusive areas of impermeable rock.
- Intermittent drainage is shown by disappearing streams.
- Many limestone areas have also been affected by glaciation.
- The lack of vegetation is due to thin upland soils.
- Caves will normally be underground but may be marked on the surface.
- Sink holes, swallow holes and potholes can all look quite similar but may be accompanied by text on the map.

Names may give clues to identify features

Lip at front of corrie

Arête forming between two corries

Steep corrie backwall

Corrie lochan

U-shaped valley with steep slopes

Hanging valleys are often difficult to identify but may occur where streams flow out of corries

Scree slopes formed by physical weathering

Glaciated uplands

- Other features to look for include pyramidal peaks, misfit streams and waterfalls from hanging valleys.
- Truncated spurs may be identifiable between hanging valleys.
- Corries predominantly face north.

Beaches are evidence of deposition

Stacks and stumps

Different rock strengths form headlands and bays

Cliff faces

A 'Down' shows this is chalk landscape in southeast England

Presence of groynes may show problems with erosion

Coastlines of erosion and deposition

- Features of both erosion and deposition are often shown on the same map extract.
- Other erosional features to look for include wave-cut notches and caves.
- Other depositional features to look for include spits, tombolos, bars and lagoons.
- Cliffs may be identifiable by very close contour patterns.
- Coves may indicate alternating bands of weak and resistant rock running parallel to the coast.

To describe relief on an Ordnance Survey map, look at the heights (using spot heights and contours), slopes and landforms and try to identify possible rock types.

DON'T FORGET

Knowing how to identify features from Ordnance Survey maps (using six-figure grid references) could be only half the story – you may then have to describe and explain their formation using detailed annotated diagrams.

LET'S THINK ABOUT THIS

It's best to learn about the formation of as many features as you can but, if you are rushed, make sure you know corries, stalagmites and stalactites and stacks and stumps – you can probably write more detail about these than anything else.

SOILS

The **biosphere** is the zone of the Earth where plants and animals live. The parts studied for Higher Geography are its soils and vegetation.

SOIL CONTENT

A **soil** is the top layer of the Earth's surface. It consists of:

- **mineral matter** from weathered **parent material**. This is known as **regolith** (rock, alluvium or glacial till) and is normally the largest part of the soil by volume.

- **organic material** from decaying plant matter and dead organisms broken down by **soil fauna** (or **soil biota**) – the micro-organisms and animals including worms and moles that live in the soil

- **air** and **water** found in the **voids** between soil particles. The quantities vary, with saturated soils having much more water and less air than well-drained soils.

THE SOIL SYSTEM

Some of the main inputs, processes and outputs of the soil system are shown below.

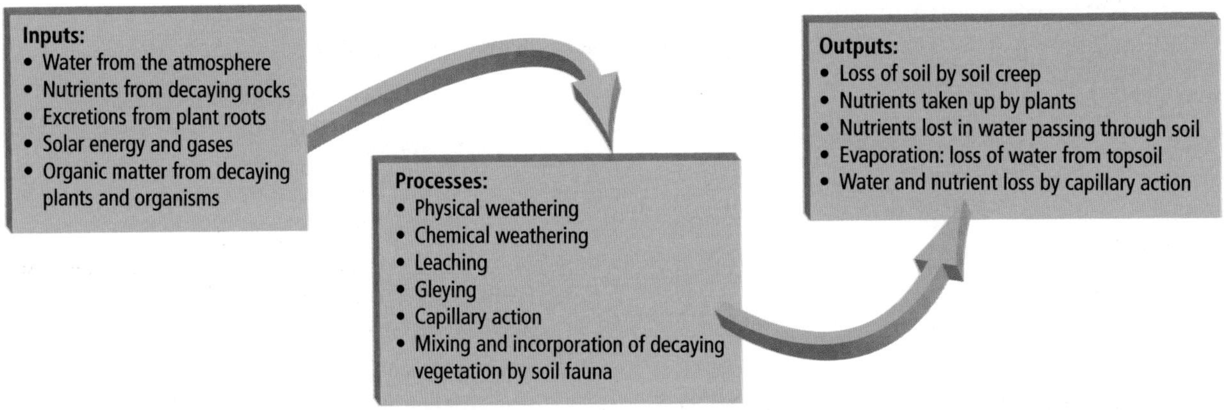

Inputs:
- Water from the atmosphere
- Nutrients from decaying rocks
- Excretions from plant roots
- Solar energy and gases
- Organic matter from decaying plants and organisms

Processes:
- Physical weathering
- Chemical weathering
- Leaching
- Gleying
- Capillary action
- Mixing and incorporation of decaying vegetation by soil fauna

Outputs:
- Loss of soil by soil creep
- Nutrients taken up by plants
- Nutrients lost in water passing through soil
- Evaporation: loss of water from topsoil
- Water and nutrient loss by capillary action

Aspects of the soil system

SOIL FORMATION FACTORS

<u>Humification</u> is the process by which leaf litter is broken down into an organic material known as **humus**. **Mor** humus is acidic and forms underneath coniferous forest. **Mull** humus is richer, nearer to chemically neutral and is found under deciduous woodland, while **modor** is humus with characteristics lying between the other two.

<u>Climate</u> determines how long it will take for organic matter to decay. Warmer temperatures mean more **decomposition** and abundant soil fauna, which encourages greater mixing of the humus into the soil below. Precipitation influences the passage of nutrients through the soil; and, in areas with high rainfall or snowmelt, leaching will remove minerals and humus. In warmer areas, where evaporation is greater than precipitation, nutrients are drawn upwards through the soil by **capillary action**. Climate also influences the length of the growing season and vegetation type.

<u>Eluviation</u> and <u>illuviation</u> are important processes related to leaching and the movement of water through the soil. **Eluviation** is where soluble minerals and humus are washed out of the A horizon, while **illuviation** is the washing in and depositing of soluble material in the B horizon below.

contd

SOIL FORMATION FACTORS contd

Relief influences drainage. Water will flow through soils found on the side of a hill and accumulate at the foot of the slope. Relief also affects local climate. South-facing slopes are warmer, as they receive more sunlight; and temperatures will decrease with increasing altitude.

Parent material affects the speed of soil formation. Hard rocks form thin soils, as they take a long time to break down. Parent material also influences the chemical composition of a soil, its colour and texture – whether it is made up of silt, sand or clay particles.

Soil fauna (biota) carry out the chemical decomposition that forms humus and mix the organic and mineral material together. Rapid mixing by ants, worms, millipedes, mites and springtails, often in milder climates, forms mull humus. Mites and springtails are more important in colder climates, where the slow breakdown of material produces mor humus.

Time is crucial in soil development – it may take a hundred years to form a centimetre of soil, so older soils tend to be thicker. Soils in equatorial areas unaffected by dramatic events such as glaciation are often tens of metres thick. Soils in the UK are rarely deeper than a metre and a half.

Human activity has had a big impact on soil formation. The removal of Scottish woodlands by prehistoric people, for instance, led to soil erosion and the development of heather moorland.

> **DON'T FORGET**
>
> Many soil formation factors are interrelated. For instance, climate influences weathering, humification and soil biota, while relief affects drainage and microclimate.

> **DON'T FORGET**
>
> Make sure you know the difference between soil processes (the ways in which a soil is formed) and soil properties (the characteristics of the soil itself).

THE SOIL PROFILE

Soils are divided into a series of different layers known as **horizons**.

The **Ao** horizon is the uppermost layer, consisting of organic matter in various stages of decomposition:

- the **L** or **Leaf Litter** layer of undecomposed pine cones and needles, grass and leaves that have fallen on top of the soil

- the **F** or **Fermentation** layer, showing the first stages of decomposition

- the **H** or **Humus** layer of completely decomposed plant and animal matter.

The main nutrient-rich layer of **topsoil** is known as the **A** horizon and consists of a mixture of mineral matter and organic matter introduced from above.

The **B** horizon, or **subsoil**, has a coarser texture and contains more mineral matter from weathered parent material. Soluble organic matter may also be washed in from above by **leaching**.

The **C** horizon is the regolith (weathered parent material), which contributes mineral matter to the upper horizons.

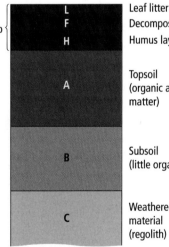

A model soil profile

 Good images and explanation of soils and soil terminology from Learning and Teaching Scotland can be found at http://www.ltscotland.org.uk/resources/g/ nqresource_tcm4465917.asp?strReferringChannel=nq

LET'S THINK ABOUT THIS

Questions on soil formation and the processes involved may be general in nature but may also refer to the specific soil types described in the next section.

SOIL PROFILES

SOIL CLASSIFICATION

Although there are many different soils, they can be divided into three types:

1. **Zonal** soils form across specific climate and vegetation zones. They include **podzols** and **brown earths**.

2. **Intrazonal** soils are more strongly influenced by parent material and can form across different climatic zones. **Gley** soils are an intrazonal example.

3. **Azonal** soils are immature and show little development of horizons. They include alluvial and volcanic soils that form independently of climate or vegetation.

SOIL CATENAS

Although zonal soils form across broad climatic areas, there can be widespread variations on a local level. A soil **catena** is a sequence of different soil profiles that develop down a slope. Podzols are found in the colder climate near to the top of the hill, brown earths on more freely drained land further down and waterlogged gleys in the valley bottom.

 More on the soil profiles shown can be found at http://www.macaulay.ac.uk/ soilquality/Soils%20and%20their%20main%20characteristics.pdf

EXAMPLES OF ZONAL SOIL PROFILES

Mild climate, moderate rainfall

precipitation ≥ evaporation
deciduous woodland

6–8 months growing season

layer of undergrowth

abundant soil biota (worms, insects, rodents)

L fresh leaf litter
F decomposing layer
H mild mull humus
} A₀

dark brown humus-enriched
A horizon

some limited leaching

long tree roots

gradual merging boundary

capillary action in warmer months

lighter brown
B horizon

deep roots help break up parent material and take nutrients up through the soil

Brown earth soil profile

weathered parent material

Brown earths

Brown earths are found under deciduous woodland in a band stretching across Western Europe, the east of Asia and the eastern side of the USA. Trees lose their leaves in autumn, and the thick undergrowth contributes to the dense leaf litter.

The mild climate encourages quick decomposition and the formation of thick, mull humus. Rates of leaching are low, as evaporation cancels out precipitation. Capillary action may bring minerals and nutrients up through the soil in warmer months. Worms, insects and larger mammals such as moles mix leaf litter and humus, and deep tree roots absorb nutrients, bringing them up through the soil. The intense mixing means that there is little colour difference between horizons.

Soils are quite deep and fertile and are often agriculturally productive once the forest cover has been removed.

contd

EXAMPLES OF ZONAL SOIL PROFILES contd

Podzols

Podzols are associated with the areas of coniferous forest that stretch across northern Europe, Asia and the Americas. Winters are long and cold and summers are cool. Although precipitation is generally low, there is little evapotranspiration, so moisture moves down through the soil. The resulting leaching takes organic matter and clay out of the A horizon, leaving it a pale ash-grey colour, and deposits it in the B horizon. Accumulation of iron may result in the formation of a hard, rust-coloured **iron pan**. This prevents water from passing through the soil and encourages waterlogging. Clay compounds in the B horizon give it a reddish-brown colour.

There is little mixing of organic matter, as the cold climate discourages soil fauna and bacteria and the leaf litter is hard and decomposes very slowly. The **mor** humus layer that develops is thin and acidic.

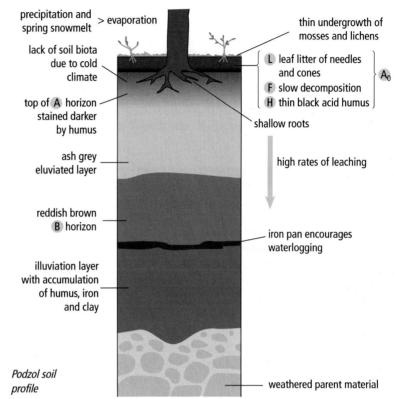

Coniferous forest, short cool summers

precipitation and spring snowmelt > evaporation

lack of soil biota due to cold climate

top of Ⓐ horizon stained darker by humus

ash grey eluviated layer

reddish brown Ⓑ horizon

illuviation layer with accumulation of humus, iron and clay

Podzol soil profile

thin undergrowth of mosses and lichens

Ⓛ leaf litter of needles and cones
Ⓕ slow decomposition
Ⓗ thin black acid humus } A₀

shallow roots

high rates of leaching

iron pan encourages waterlogging

weathered parent material

Gley soils

Gleys are found in temporarily or permanently waterlogged areas where the voids become filled with water and lose oxygen. In these **anaerobic** conditions, leaf litter decay is slow and there is little soil fauna, so a large amount of organic matter accumulates.

In anaerobic conditions, iron compounds in the soil are chemically reduced and change colour from red or brown to blue or grey. If the gley soil dries out, this process can be reversed, and the heavily waterlogged blue-grey B horizon takes on a **mottled** appearance.

Waterlogging can take place at the foot of a slope where water accumulates, on flood plains or where soils form on an impermeable parent material.

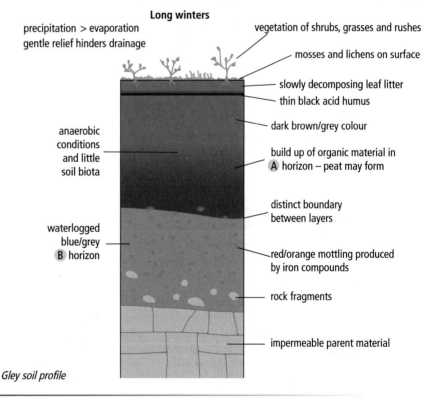

Long winters

precipitation > evaporation
gentle relief hinders drainage

anaerobic conditions and little soil biota

waterlogged blue/grey Ⓑ horizon

Gley soil profile

vegetation of shrubs, grasses and rushes

mosses and lichens on surface

slowly decomposing leaf litter

thin black acid humus

dark brown/grey colour

build up of organic material in Ⓐ horizon – peat may form

distinct boundary between layers

red/orange mottling produced by iron compounds

rock fragments

impermeable parent material

⚙ LET'S THINK ABOUT THIS

It is quite likely you will be given a choice of soil profiles to describe in an exam – so, if you can't learn all three, make sure you know two in detail and have at least a little knowledge of the one you're less sure of.

PLANT SUCCESSION

The biosphere is made up of many different **ecosystems** – communities of plants and animals found in a specific **habitat**. Soils, vegetation, climate and animals are all interrelated within an ecosystem. They exist at many different levels from a single tree, through a deciduous woodland, to the whole planet.

Studies of **plant succession** are concerned with the sequence of stages that lead to the development of ecosystems known as **climax vegetation**.

SERAL SUCCESSION

DON'T FORGET

It is important to have a detailed knowledge of the stages that lead to climatic climax vegetation – explaining how it occurs is a fairly common exam question.

The progression to climax vegetation involves a number of different **stages** with the growth of increasingly complex plants. These stages are known as **seres**.

The first stage is the **prisere**. Basic plants such as lichens and algae colonise areas of bare rock, sand or other **substrate**, die and add organic matter while also weathering the underlying parent material. Mosses may then colonise, and a basic soil begins to form, bound by the roots of the first plants.

Gradually, grasses may replace the **pioneer stage** plant species and, as physical conditions such as soil, shade and light change, the succession proceeds through a number of **building stages**. Earlier plants will be replaced by shrubs and smaller trees such as willow. In many cases, woodland such as Scots pine or oak will eventually become dominant as the climax vegetation. This final ecosystem is in equilibrium with the climate and soils of the area in which it forms and is known as the **climatic climax vegetation**.

There are many kinds of plant succession: **lithoseres** develop on bare rock, **hydroseres** near water and **haloseres** on salt marsh. **Sand dune succession** is known as a **psammosere**.

DON'T FORGET

The psammosere is the only kind of plant succession that is examined.

Biomass

The **biomass** is the total weight of living matter in a certain area, and estimates of biomass can be used to compare ecosystems. It includes everything living below ground so, in the case of plant biomass, comprises roots as well as branches, leaves and everything visible above the soil. Animal biomass is also included but is small compared to that of plants. Biomass increases across seral stages, being at a minimum in the pioneer stage and reaching a maximum in the climatic climax vegetation.

SAND DUNE SUCCESSION

Sand dune systems are dynamic and in constant motion as sand is blown inland by the wind. If the blown sand is trapped by an obstacle (e.g. a piece of driftwood, seaweed or a tin can), then it provides a more stable base on which plants can grow.

The trapped sand builds up as a small **embryo dune** near to the shore line. Pioneer plants that **colonise** here need to be resistant to immersion in high tides and able to withstand high winds. The sand is poor in nutrients and is very dry because of rapid drainage. Plants that can cope with these dry conditions are called **xerophytic**. Fragments of shells found on the shore line also make the soil alkaline.

Plants growing on the embryo dunes trap more sand and, as the depth of the dune increases, they become **fore dunes**. Growing conditions are still very tough but, as leaf litter is added to the soil, a basic humus begins to develop. Plants found on the embryo and fore dunes include **sea rocket, sea twitch, sea couch, lyme grass** and **sandwort**.

contd

SAND DUNE SUCCESSION contd

Above the high tide level, **marram grass** becomes the dominant plant. It has long and widespread stems (called **rhizomes**) which spread underground vertically and horizontally, binding the sand and reaching water. The leaves curl up in order to retain moisture. Marram grows quickly – up to one metre a year – to keep above the sand that piles up to form the highest **mobile** dune range of **yellow dunes**.

As the depth of sand increases, the dunes shelter the hollows further inland, and other plant species can become established on the **fixed** or **grey dunes** further back from the shore. Plants growing here include **sea holly, dandelion, creeping fescue, sand sedge** and **bird's-foot trefoil**.

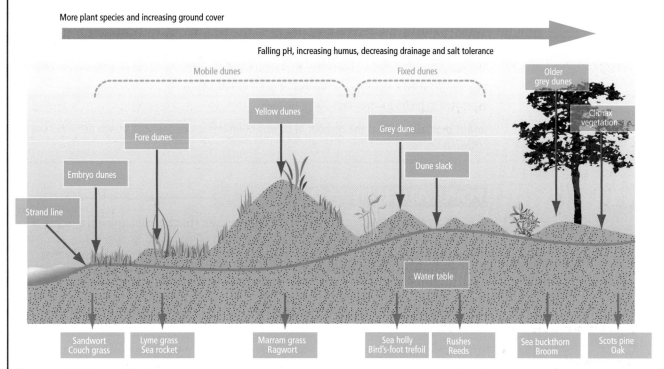

Plant succession in a sand dune system

The older grey dunes have more established plants and bushes including **sea buckthorn, broom** and **heather**. The humus formed on these dunes is more acidic, as all traces of alkaline shells have been leached out.

Between the ranks of grey dunes are lower-lying hollows called **dune slacks** where the water table reaches the surface. **Hydrophytic** marsh plants are dominant, including **rushes** and **creeping willow**. Deeper and wetter soils inland may also become home to **ash** and **hawthorn**.

The last stage in the psammosere furthest back from the beach is the climatic climax vegetation. In Scotland, this would normally be **Scots pine** and **birch** or **oak**. In many areas, however, human activity and land management has intervened and a **plagioclimax** has formed instead.

Psammoseres are very fragile, especially in their early stages of growth, because they can be easily destroyed by the wind or washed away by high tides.

DON'T FORGET

Make sure you can name at least one plant species in each seral stage, which dune they occupy and what conditions are necessary for them to grow.

 A great PowerPoint with pictures of sand dune succession can be found at http://www.slideshare.net/ValVannet/biosphere-sand-dune-succession-32557

LET'S THINK ABOUT THIS

Questions about sand dune succession have recently been as common as those on soils and soil profiles. While they may seem daunting, you can pick up fairly easy marks if you learn the processes involved and the names of at least one plant in each stage of the succession.

POPULATION STRUCTURE

POPULATION TERMS

Demography is the study of population.

Crude birth rate (CBR) is the number of people born each year per 1000 head of population. CBR takes no account of the age and sex structure of a population. An alternative measure of fertility is the **general fertility rate** – the number of live births per 1000 women aged 15 to 44.

Crude death rate (CDR) is the number of people who die each year per 1000 head of population.

Natural increase, calculated as birth rate minus death rate, is the increase in population per 1000 people. It can also be expressed as a percentage.

Life expectancy is the average number of years somebody will live for. Globally, this is estimated to be 65 for men and 70 for women.

Infant mortality rate is the number of deaths of infants under one year of age per 1000 live births.

POPULATION PYRAMIDS

Population pyramids are used to show the **age and sex structure** of a country, a region or other administrative area. A pyramid shows the numbers of males and females in five-year bands starting with infants (aged 0 to 4 years) and continuing to 80 years and over. Normally they show the percentage of males and females in each **cohort**, but sometimes absolute numbers are used instead.

Countries at different stages of development typically show different characteristics in their population pyramids.

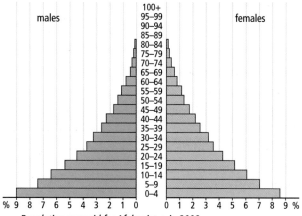

Population pyramid for Afghanistan in 2008

Population pyramid for Germany in 2008

- a **wide base** as a result of a high CBR

- a **narrow peak** showing a relatively low life expectancy

- big decreases upwards from one age cohort to the next indicating a high CDR and, near the base, a high infant mortality rate

- a high percentage of the population aged between 0 and 14 (***this structure is indicative of a very poor developing country***)

- a **narrow base** due to a low and decreasing CBR

- a **blunt peak** and **straight sides** gradually tapering upwards to show low CDRs and a long life expectancy

- a high percentage of the population aged 65 and over reflecting an **ageing population** (***this structure is indicative of a relatively rich developed country***)

Both pyramids show the longer life expectancy of women that exists in nearly all countries.

contd

POPULATION PYRAMIDS contd

The dependency ratio

The **dependent population** refers to those people who do not work, so includes **young dependants** aged between 0 and 14 and **old dependants** aged over 65 who have usually retired.

The **economically active** or **working population** is aged between 15 and 64.

The **dependency ratio** is calculated as:

$$\frac{\text{\% young dependants} + \text{\% old dependants}}{\text{\% economically active}}$$

A high dependency ratio means there are fewer people in jobs making money and paying taxes to provide for those too young or too old to work. The analysis of dependency ratios has implications for planning in education and health and in calculating the tax burden. Young dependants will need money spent on education, whereas old dependants may need more expensive health services, especially as people are living longer and medical advances are increasingly costly.

PROBLEMS CAUSED BY DIFFERENT POPULATION STRUCTURES

Developed countries

An increasingly ageing population causes **stagnation** and, sometimes, **population decline**. Having more old dependants increases the burden on the economically active because of the increasing costs and strain on pension schemes, health care, home care services and sheltered housing and the need for greater public transport.

Having fewer births and fewer young dependants may result in the closure of schools and maternity wards.

It may be necessary to raise the retirement age to compensate for the lack of economically active people or to raise taxes for those still in work. Encouraging immigration is a solution to increase the number of people in work but can bring other social problems.

Developing countries

A high percentage of young dependants means more money is spent on child health care and education. Schools cannot cope, and young girls are less likely to be educated, so birth rates remain high. Over time, more young people move into adulthood and become parents. Many are unemployed because the number of jobs cannot keep pace with the population increase.

A high and increasing young dependent population increases the demand on food, housing and other resources.

The number of old dependants also increases with time, leading to more strain on scarce services and poor infrastructure.

An increasing problem in sub-Saharan Africa is the number of sexually active people dying from AIDS. This is causing a big bulge in the younger age cohorts, an increase in the dependency ratio and a strain on poorly funded health and education services.

 Population pyramids for almost every country can be built at http://www.census.gov/ipc/www/idb/

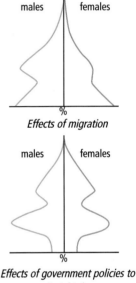

males   females

Effects of migration

males   females

Effects of government policies to limit births

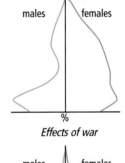

males   females

Effects of war

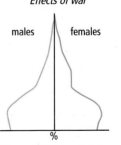

males   females

Effects of AIDS in sub-Saharan Africa

LET'S THINK ABOUT THIS

Population pyramids provide a quick visual comparison between population structures of different countries. Make sure you can interpret them, can identify specific events and can consider the implications of certain structures for the future of the countries concerned.

DEMOGRAPHIC TRANSITION

MODELLING DEMOGRAPHIC TRANSITION

The **demographic transition model** attempts to show how population changes over time. It has five different stages and can be used to explain how CBRs and CDRs respond to different economic and social situations. The stages can also be drawn as population pyramids.

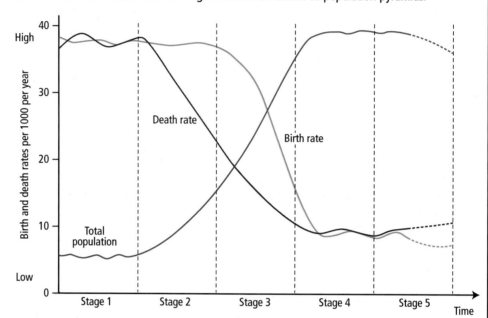

The demographic transition model

Stage 1: High fluctuating

- Death rates are high because of wars, famine, plagues and epidemics, a lack of clean water and very basic medical care.
- Birth rates are high because of a lack of birth control, low marriage age and because children work, mainly in agriculture, to add to the family income.
- Life expectancy and average age are low.
- Population growth is slow or stagnant.

Scotland was in this stage pre-1760 but, today, only a few remote tribes in areas such as the Amazon and New Guinea show these characteristics.

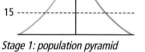
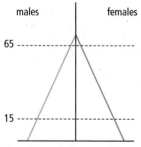

Stage 1: population pyramid

Stage 2: Early expanding

- Death rates drop because of: surgical advances, the increased availability of medical supplies and the introduction of vaccines for diseases such as smallpox; improved nutrition, sanitation and water supplies; rising wages leading to better personal hygiene.
- Birth rates remain high because of religious factors, the importance given to men with large families and the low status of women in society.
- Life expectancy increases, infant mortality decreases.
- Population growth is very rapid.

Scotland was at this stage between 1760 and 1870. Today, some very poor developing countries such as Afghanistan and Sierra Leone are in this stage, but most have moved on because of improving health care.

Stage 2: population pyramid

 Detailed information on Scotland's population structure and characteristics can be found at http://www.scrol.gov.uk/scrol/common/home.jsp

MODELLING DEMOGRAPHIC TRANSITION contd

Stage 3: Late expanding

- Death rates continue to drop with further improvements in sanitation, health care and medical facilities.
- Birth rates show a huge drop due to: the adoption of family planning and the development and spread of contraception; declining infant mortality rates; the desire for consumer goods, combined with the increased costs of keeping children, leading to smaller families; mechanisation meaning fewer labourers needed for agriculture and factory work. Better educational opportunities for women and higher female literacy rates also decrease the desire for large families.
- Life expectancy continues to increase.
- Population growth is still rapid but slowing.

Scotland was at stage 3 from 1870 to 1950. Many developing countries, including India, Brazil and Mexico, are now at this stage.

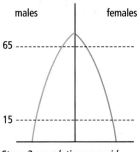
Stage 3: population pyramid

Stage 4: Low stationary

- Birth rates have decreased to roughly the same level as death rates but continue to fluctuate in response to economic conditions. Very effective birth control enables people to limit the number of children they want, and the average marriage age increases as women pursue careers before starting families. Migration of young, fit individuals also contributes to the fall in birth rates.
- Population is high but steady.

Scotland entered this stage around 1950. Some richer developing countries, e.g. Argentina, will soon be at stage 4.

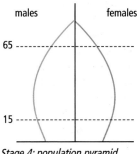
Stage 4: population pyramid

Stage 5: Declining population

- Death rate shows a possible rise because a greater proportion of the population is elderly.
- Birth rate is low and decreasing as people delay child-rearing or make a lifestyle or economic choice to have only one child.
- Population is decreasing.

So far, only a few highly developed countries with very low fertility rates such as Italy, Germany and Sweden are in this stage. Scotland may enter stage 5 soon.

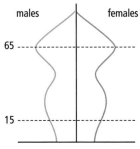
Stage 5: population pyramid

NATALIST POLICIES

Governments often try to influence the birth rate in their countries through the introduction of anti-natalist or pro-natalist policies.

Anti-natalist policies discourage childbirth by removing financial benefits or imposing financial penalties on those couples who have more than one child. Massive contraception campaigns and adverts showing the benefits of small families may be launched. Policies can be very forceful and involve compulsory abortions and sterilisation and, in China for example, public denunciation and huge social pressure to have only one child.

Pro-natalist policies encourage larger families, usually with rewards or financial benefits. In Nazi Germany, medals were given to women with lots of children. More recently, in the UK, increased child-benefit payments and tax credits have been introduced. Flexible working hours for parents and longer periods of maternity and paternity leave are also common in developed countries such as Sweden and Norway.

DON'T FORGET

China's one-child policy is an excellent example of anti-natalist policies at work. Up-to-date examples of pro-natalist policies, especially from the UK, are also useful.

LET'S THINK ABOUT THIS

The demographic transition model has been criticised because it is based on European history. Regional variations in poorer continents and differences between urban and rural areas within countries also need to be considered. Make sure you understand how the model relates to any individual countries you use as case studies.

CENSUS AND MIGRATION

COLLECTING DEMOGRAPHIC DATA

In many countries, including the UK, population information is collected through **civil registration**. All births, marriages and deaths have to be registered, usually at local council offices. **Population surveys** looking at social and economic data are carried out at national, international and global levels by many organisations, including the Scottish Government and the United Nations.

Censuses

Many countries use a detailed form called a **census** to get information on the social and economic make-up of their population.

The data collected includes details about each member of a household including their age, sex, religion, health, job and means of transport to work, education, birthplace and, in Scotland, knowledge of Gaelic.

In the UK, a census is conducted every 10 years at the start of the decade (e.g. 2011). The whole country is divided into small areas called **enumeration districts**, and completed household forms are collected by an **enumerator**. The results are analysed to look at population change and future demand for services. They are then published and used by government, local planners and organisations interested in health care, education, employment, housing, transport and business.

 Look through the results of the last Scottish census at www.scrol.gov.uk

Problems in collecting demographic data

Census data is generally reliable in developed countries, but this is not always the case. In the UK, many people failed to complete the 1991 census because of worries over the poll tax.

In developing countries, there are a number of reasons why collecting demographic data is difficult:

- Running a census is expensive for cash-strapped governments with other priorities.
- Nomadic people and illegal immigrants (including Mexicans living in the USA) are often not taken into account.
- Countries with many official languages have to spend more money translating forms and employing enumerators. Nigeria, for instance, has six major languages and hundreds of unofficial ones.
- Civil war, refugees and conflict cause problems, and political differences may lead to officials submitting phoney figures to favour certain ethnic groups.
- High adult illiteracy means that people are unable to fill in the forms.
- Poor infrastructure and difficult terrain hamper data collection.
- Social and religious reasons may cause under-recording of females – in China, under the one-child policy, many female births went unrecorded.
- Rapid migration to urban slums can mean that large numbers of people are forgotten.

DON'T FORGET

When thinking about the difficulties in collecting demographic data, relate them to specific countries. For example, Brazil's widely dispersed population, mountains, wetlands and large areas of rainforest make complete coverage a real headache for the authorities.

MIGRATION

When people move from one place to another on a permanent or semi-permanent basis, it is known as **migration**. This can be for a matter of months, for years or on a **seasonal** basis (e.g. fruit-pickers from the continent coming to work in Kent).

Within developing countries, the most common movement is from rural to urban areas. **Push** factors force people away from the countryside and into the cities, where they are attracted by **pull** factors.

contd

MIGRATION contd

- Loss of farmland through land reform and population pressure
- Mechanisation leading to agricultural job losses
- Low wages from agricultural employment
- Low standard of living
- Poor sanitation, water supply and rural infrastructure
- Poor rural services such as health care and education
- Drought leading to loss of crops
- Land degradation and decreasing productivity of agricultural land

Push factors affecting migration

- Job opportunities in manufacturing and service-sector industries
- Better educational opportunities
- Better health care, doctors and hospitals
- The "bright lights" of the big city
- Higher wages than in rural employment
- Extended family already moved to the city
- Possibility of better housing, sanitation and a higher standard of living
- Access to amenities, shops and entertainment for young people
- Many of the urban pull factors may, in reality, prove to be hard to find or expensive and out of the reach of rural migrants

Pull factors affecting migration

Migration between countries is often **voluntary** as migrants move to find better-paid jobs. **International** migration can be **forced** when people move because of natural disasters such as drought or earthquake or because of war or religious or political persecution. Under such circumstances, the migrants are usually referred to as **refugees**.

Barriers to migration

There may be a number of reasons why people, who appear to have little to stop them, do not migrate. These include laws against immigration in receiving countries, the costs involved, racial problems and a lack of housing, education and employment in the countries where they want to go.

> **DON'T FORGET**
>
> In many developed countries, urban–rural movement is now the dominant form of migration – a process known as **counter-urbanisation**.

ADVANTAGES AND DISADVANTAGES OF MIGRATION

Both the areas losing and receiving migrants get benefits and suffer problems from the movement – these are usually interrelated.

- Migrants are often young, economically active, well-educated males, so the receiving area gets a dynamic part of the population.
- Rural communities in the losing areas may end up with a sex and skills imbalance and an increasingly ageing population.
- Labour shortages are filled in the receiving countries, and declining populations may be reinvigorated.
- Emigrants often send money home to their families, and, when they return, the skills they have learned will enable them to get better jobs.
- Immigration causes pressure on services in the receiving areas and competition for houses, jobs, education and health care. This can lead to racial tensions and, occasionally, violence and rioting.
- There is reduced pressure on scarce resources and food in the country losing migrants.
- Immigrants may increase the cultural diversity of the receiving country.
- Particularly in developing countries, migrants may lead to the spread of shanty towns around urban areas.

> **DON'T FORGET**
>
> You need to know case studies of different types of migration and the effects on the countries and people involved: international migrations such as Mexicans to the USA and Poles to the UK; national migrations from the city to the country in the UK or from the countryside to the city in Nigeria; and forced movements such as the increasing number of refugees from the Darfur region of Sudan.

Read the rural migration news from California at http://migration.ucdavis.edu/

LET'S THINK ABOUT THIS

In the exam, Population questions often refer to graphs or diagrams – but remember, many of the topics are interlinked, so, for example, a question interpreting population pyramids may have reference to migration in the answer.

FARMING SYSTEMS AND SHIFTING CULTIVATION

RURAL TERMS

Arable farming is concerned with growing crops, while **pastoral** farmers look after livestock such as sheep and cows.

Subsistence farmers produce food to feed themselves and their families, with, if possible, a little surplus they can sell. **Commercial** farming is carried out to produce goods for sale in markets.

Intensive farming usually takes place on a small scale and involves high inputs of either capital (in the case of market gardening) or labour (with rice grown in padi fields). **Extensive** farming is carried out over large areas and often involves high inputs of technology.

THE FARMING SYSTEM

All farming systems start with **inputs** such as soils, climate, labour, seeds and fertilisers. To get to the **outputs** (crops or livestock), a number of **processes** or activities are carried out. A systems diagram for shifting cultivation is shown below.

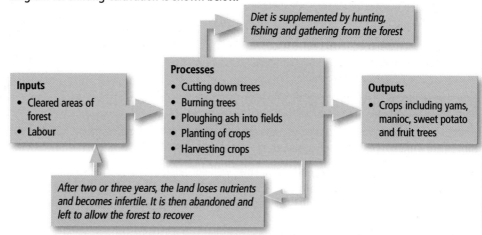

Simplified farming systems diagram for shifting cultivation

The available inputs to a system will determine the type of farming that takes place in a certain area.

Aspects of three farming systems – **shifting cultivation, commercial arable farming** and **intensive peasant farming** – are studied for the Higher exam:

- the landscapes associated with the system
- the main features of, and methods involved in, each type of farming system
- the associated population density
- farming changes since 1950 and the impacts on people and landscapes
- the benefits and problems brought about by changes in farming.

Shifting cultivation

An example of **subsistence agriculture**, shifting cultivation takes place in the rainforest areas of South America and Africa and in some parts of south-east Asia. Large areas of land are needed to support individual tribes and communities, who cultivate small plots for a few years and then move on as soil fertility declines. Population density remains low.

The following description relates to a typical case study of shifting cultivation by Amazonian tribes in Ecuador and Brazil.

contd

THE FARMING SYSTEM contd

Features of the farming landscape

A large area of land is cleared in the rainforest, normally near a river which will serve as the main method of transport. Houses are built, along with communal living areas for food preparation and village meetings.

Farming plots in the forest surrounding the settlement are cleared by hand. Axes are used to fell the larger trees, and bushes are cleared with machetes or other basic tools. The vegetation is then burned to add **nutrient-rich ash** to the **infertile** tropical soils.

Most of the digging is done by women, who then plant a wide variety of crops, including banana, manioc, sweet potatoes, beans, yams, taroes, groundnuts, pineapples and other fruits as well as various spices, and poisonous plants that will be used in fishing and hunting.

The land is cultivated for two to three years, by which time the intense tropical rain has washed the nutrients out of the soil. The land is then left to recover as secondary forest (a process which takes up to 30 years), and another plot of land within easy reach of the settlement is cultivated.

Hunting and gathering add fish, bushmeat, fruits, nuts and honey to the already varied diet.

⌇ DON'T FORGET

You may be given a diagram of a landscape of shifting cultivation, so make sure you know and can describe the important features.

Recent changes

Shifting cultivation in the Amazon region is under threat from increasing population density and development projects.

The shifting cultivation landscape

- Many migrants have started farming the land and reduced the area available to shifting cultivators. Newer farming methods are often unsustainable, and the immigrants use up increasingly larger areas trying to grow crops. Since the 1970s, for instance, the Brazilian government has encouraged migration to the Amazon region to reduce population pressure in the country's already crowded cities.
- Large-scale cattle ranching is taking place to provide South America's increasing population with a source of beef.
- Mineral exploitation and the search for oil is having an increasing impact. It takes up large areas of land – and the chemicals used, such as mercury, poison rivers, kill fish and trees and end up in drinking-water sources used by the indigenous population.
- Large-scale hydro-electric projects are flooding land used by shifting cultivators.
- Roads are being cut through large areas of rainforest to open up the land for the above activities.

The impacts of recent changes

As the land available for shifting cultivation decreases, indigenous people are often forced into settled agriculture using fertilisers and other modern (often environmentally unsustainable) methods to maintain soil fertility.

Some tribes have been removed from their land by force. Murder and violence are increasingly common as more people compete to use limited resources.

Indigenous people who leave the land forget their traditional ways of life and add to population problems in Brazil's already overcrowded urban areas.

Benefits brought about by the changes may include:

- increased job opportunities for indigenous people in mining, logging and cattle ranching
- greater services, such as medical care and education, available to previously isolated and primitive tribes.

 Information and video clips on changing lifestyles in the Amazon can be found at http://www.bbc.co.uk/amazon/

LET'S THINK ABOUT THIS

When answering questions about how recent changes have affected shifting cultivators, there is a tendency to dwell on the negative impacts – examiners want to know that you understand the positive features of change as well.

COMMERCIAL ARABLE FARMING

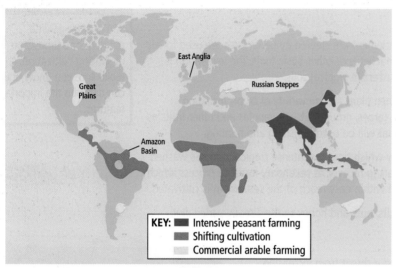

Worldwide distribution of selected farming systems

KEY:
- Intensive peasant farming
- Shifting cultivation
- Commercial arable farming

In extensive agriculture, a relatively small amount of produce (**yield**) is obtained from a large area of land. Both crops and animals can be farmed extensively. Examples include hill sheep farming in Scotland, beef cattle ranching in Argentina, arable farming in Norfolk and the cultivation of wheat on the North American prairies.

The American Midwest is the most common case study for commercial arable farming, but other examples include the Russian steppes and East Anglia. Although there will be specific differences between the case studies, the main characteristics of the farming system remain the same.

FARMING ON THE GREAT PLAINS

The Great Plains of North America stretch from southern Canada to the Mexican border and, west to east, from the Rocky Mountains to the central lowlands of the USA. Population density is low, but farming is highly **mechanised** and produces large amounts of wheat for both the US domestic and export markets worldwide.

Components of the farming system are as follows:

Inputs

- Relief – the land is mainly flat, and this has encouraged the introduction of new farming technologies and machinery.
- Climate – very variable over such a large area but with sufficient sunshine for at least one crop of wheat each year.
- Water – variable rain and snowfall is supplemented with spring meltwater at the start of the growing season or by irrigation water from underground aquifers.
- Soils – deep, dark **chernozems** (black earths) which are fertile and rich in humus.

- Capital – extensive farming is **capital intensive**, and large amounts of money are needed for: seeds; machinery such as combine harvesters; fertilisers to maintain soil fertility; pesticides; irrigation; transport for the long distances to markets.
- Labour – demands are generally low but increase seasonally, so **contractors** are employed, especially during harvesting.
- Political influence – **grants** and **subsidies** paid for certain types of crops are important in influencing decisions made by farmers; other factors include international agreements on tariffs and **protectionist** trading policies.

Processes

- Investment – in new technologies and improved quality of seeds, fertilisers and other chemicals
- Ploughing, fertilising and planting
- Harvesting – often by specialist teams contracted for short periods of time who travel the prairies with their own equipment

- Storage and processing – requiring investment in large buildings
- Transportation – either by rail or road to ports and markets.

Outputs

- Spring wheat and, if possible, a winter wheat crop
- Occasionally barley, sugar beet and fodder crops such as alfalfa
- Cattle fattened on large cattle ranches in areas where soils are less fertile.

FEATURES OF THE FARMING LANDSCAPE

The farming landscape has a very **geometric** pattern because it was originally divided into 64-hectare blocks by surveyors before the land was settled. Settlers were given an individual block, and farms increased in size as successful farmers bought up land from those who gave up when drought and other hardships forced them to leave.

Railway networks – the primary way of moving large amounts of grain – followed the block field pattern and perpetuated the geometric nature of settlement. Later, the road system was established in a similar format.

Hamlets, villages, towns and cities grew up at regular intervals in a **hierarchical** fashion along the railway network to provide services to farmers.

As more farms began to **amalgamate** in the late 20th century, farms grew massively in size. Field sizes have also increased and now take up hundreds of hectares.

The American Midwest farming landscape, with large fields and a central farmhouse or isolated settlements

RECENT CHANGES IN THE GREAT PLAINS

Rural depopulation started when **dustbowl** conditions were caused by drought in the 1930s, and has continued into the 21st century. This, combined with variable climatic conditions, has affected farming in the area.

Changes have included:

- A continued increase in farm size
- The introduction of specialised farming methods including contour ploughing, strip cultivation, shelter belt planting and other land management schemes
- Dependence on agricultural technology (including mechanisation, new seed types and chemicals) has increased capital expenditure and has had negative effects on the environment
- Increased use of irrigation leading to the depletion of important **aquifers**, some of which are already contaminated by agricultural chemicals
- Increased use of agricultural contractors, decreased permanent agricultural employment and more part-time farming
- Introduction of new crops such as sunflowers and the development of organic farming
- Land being taken out of production to allow fertility to recover
- The growth of **agribusinesses** and cooperatives

These changes have led to a fall in the rural population and diminishing services such as schools, shops and hospitals. Smaller farmers have been unable to compete with the more efficient agribusinesses, and farm labourers have been put out of work by the increased mechanisation. Land has been sold, farmsteads have been abandoned, and depression, ill-health and suicide rates have increased among the people who remain.

DON'T FORGET

It is quite common for the Rural Geography question in your exam to refer to maps, diagrams, charts and graphs. Remember to use the information provided in reference diagrams, backed up with the information from your case studies.

DON'T FORGET

Often, the main reasons behind change in farming landscapes are economic and may include government influence by changing subsidies provided to farmers.

 A PowerPoint showing commercial arable farming on the Great Plains is available at http://rhsgeography.wikispaces.com/file/view/ Extensive+Commercial+Farming+-+USA+wikispace.ppt

LET'S THINK ABOUT THIS

Remember to take the time to read the Rural question carefully – it could be about the **causes** of changes, the **changes** themselves or the **impact** of the changes, and it is crucial that you answer the question set.

INTENSIVE PEASANT FARMING

RICE FARMING IN SOUTH-EAST ASIA

In most of south-east Asia, including India, intensive peasant farming (IPF) is the dominant form of agriculture. The main crop is rice – one of the world's **staple** foods – but wheat and millet are also cultivated. Crops grow quickly and temperatures tend to be relatively constant, so, depending on rainfall or sufficient **irrigation**, up to three harvests can take place each year. Population density is high and farm sizes are small, leading to a very intensive type of agriculture.

Components of the farming system are:

Inputs

- Land – must be flat, either naturally or by artificial **terracing**
- Water – supplied by rainfall or irrigation systems from rivers and lakes (irrigated water also deposits **silt** to improve soil fertility)
- Natural fertiliser or **manure** provided by animals
- Large amounts of labour.

Processes

- Terrace construction and building of small walls (known as bunds) around fields to maintain water level
- Ploughing of rice **stubble** into soil to maintain fertility using animals or basic machinery
- Seeds initially planted in nursery beds and **transplanted** to **padi** fields filled with several centimetres of water
- Weeding, fertilising and harvesting carried out by hand
- Maintenance of fields before planting of next crop.

Outputs

- Main crop of rice with up to three harvests each year
- Other crops including wheat, maize, sorghum and millet in drier areas.

> **DON'T FORGET**
>
> Don't rely on learning only one of the selected farming systems – although normally given a choice, you may have no control over which one you will be asked to describe and explain.

FEATURES OF THE FARMING LANDSCAPE

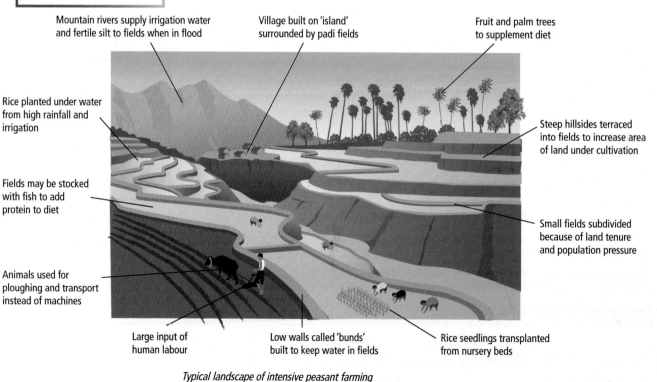

Mountain rivers supply irrigation water and fertile silt to fields when in flood

Village built on 'island' surrounded by padi fields

Fruit and palm trees to supplement diet

Rice planted under water from high rainfall and irrigation

Steep hillsides terraced into fields to increase area of land under cultivation

Fields may be stocked with fish to add protein to diet

Small fields subdivided because of land tenure and population pressure

Animals used for ploughing and transport instead of machines

Large input of human labour

Low walls called 'bunds' built to keep water in fields

Rice seedlings transplanted from nursery beds

Typical landscape of intensive peasant farming

RECENT CHANGES

Over the last 30 years, the output in areas of intensive peasant farming has increased markedly due to the introduction of new varieties of rice (see **The Green Revolution** below) but also because of changing farming methods, including:

- Planting rice directly into flooded padi fields, rather than transplanting
- Increased use of machinery such as automated ploughing as field size has increased
- Greater use of chemical fertilisers and pesticides
- Hiring machinery instead of labourers.

As farming has become increasingly intensive, farmers spend more time on the land and less on other activities such as raising cattle. They have more money to invest back into mechanisation or to buy land. Less efficient farmers have, however, lost their land and become hired labourers or have moved to towns and cities to find work in the industrial sector.

Increased use of fertilisers and pesticides has had negative effects on the environment in some areas. Chemicals have built up in the land, decreasing soil fertility, and have entered water courses, killing fish stocks that are an important addition to the diet of poorer families.

THE GREEN REVOLUTION

Breakthroughs in plant breeding in the 1960s produced **high-yielding varieties (HYVs)** of staple food crops including rice and wheat and, for many years, meant that food production kept pace with population increase in many developing countries. This **green revolution** incorporated modern farming techniques such as increased use of fertilisers and pesticides, better water control and improvements in irrigation and mechanisation.

In some countries, including India, structural changes were also needed to ensure the success of the HYVs. **Land reform schemes** produced bigger farms where mechanisation was a viable alternative to the use of human labour. Training was provided for farmers, and agricultural advisors were sent out on farm visits. Investment in rural banking and **credit agencies** meant that farmers could borrow money to improve their farms. Infrastructural improvements, including better road transport for access to markets, were also important.

DON'T FORGET

If answering a question on recent changes in agriculture, IPF and the green revolution will give you a huge amount of case study information – make sure you can relate it to the impacts on both people and the landscape.

The green revolution has meant improved wages for many farmers and more food for many countries, but there have been some drawbacks.

Pros	Cons
Yields of rice and wheat have seen a fourfold increase in some areas	HYVs need large amounts of expensive chemicals
HYVs grow quicker, so more crops are harvested each year	Increased chemical use has damaged the environment in some areas
Farmers can afford better tractors, seeds and chemicals	Irrigation schemes have led to increased **salinisation** of once fertile soils
A better diet for the rural population	
More facilities for irrigation	Rural debt levels have increased
Living standards have risen for farmers, and the **multiplier effect** has benefited other rural services	HYVs are less **drought-resistant** and more susceptible to pests and disease
New jobs in, for instance, credit agencies and chemical manufacture	Richer farmers have benefited most – some poorer farmers have lost their land and become farm labourers
Improvements in local infrastructure such as roads and health services	Unemployment and rural–urban migration have risen
	Some HYVs do not taste as good to local people

Rice Bowl Tales with case studies of IPF can be found at http://www.abc.net.au/rn/streetstories/features/rice/ while more information on the green revolution, including the latest developments, is at www.irri.org

 ## LET'S THINK ABOUT THIS

With all rural case studies, make sure you can describe the features of the landscape, the population density it supports, farming methods, recent changes and how they have affected both people and the environment.

INDUSTRIAL SYSTEMS AND INDUSTRIAL LOCATION

TYPES OF INDUSTRY

Primary industries include farming, fishing, forestry, mining and quarrying. They are involved in the **extraction** of **raw materials** straight from the earth and are also known as **extractive industries**.

Secondary industries include shipbuilding, oil-refining, chemical production, motor-vehicle manufacture and the production of iron and steel. They are concerned with the **processing** of raw materials or products from other industries and are often called **manufacturing industries**.

Tertiary industries include transport, tourism, education, shops and administration. They provide a service to the consumer and are referred to as **service industries**.

Quaternary industries are based on advances in technology and include research and development, call centres, biotechnology and information technology.

Industrial trends

Most developed countries have very few people working in primary industries but, in very poor countries like Ethiopia, where 80 per cent of the working population is involved in agriculture, this is still the dominant industrial sector.

Secondary industries add value to a country's raw materials and are very important for economic development. Most developed countries have well-established manufacturing industries. Some developing countries, including China and India, are quickly gaining wealth because of investment in secondary industries.

Service industries now dominate employment in many countries, both rich and poor. However, in developing countries, these jobs are more likely to be low-paid in areas like retailing and tourism. Developed countries will have much more service employment in banking, finance, health and education.

Increasingly important are quaternary industries – and, because they rely on new technologies such as satellites and global communication, they are less tied to specific locations and are called **footloose** industries. Call centres, for instance, can move to areas where labour is cheap, and, because of this, many have relocated from the UK to English-speaking countries like India and Jamaica.

DON'T FORGET

Although it is unlikely you will be asked about types and trends of industry in the external exam, they could crop up in your NABs and may be useful for background knowledge.

INDUSTRIAL SYSTEMS

All industry works as a system (in the same way as we saw with farming) comprising inputs, processes and outputs. A summary of the industrial system is shown below.

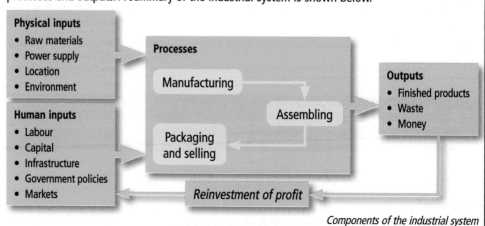

Physical inputs
- Raw materials
- Power supply
- Location
- Environment

Human inputs
- Labour
- Capital
- Infrastructure
- Government policies
- Markets

Processes

Manufacturing

Assembling

Packaging and selling

Outputs
- Finished products
- Waste
- Money

Reinvestment of profit

Components of the industrial system

INDUSTRIAL LOCATION

The factors that determine where an industry is located are varied and change over time. The reasons for industrial location in the 21st century are very different from those of the 18th century.

Reasons for the location of "traditional" industry

- **Power supply** would originally have been from water, so industry would be sited near faster-flowing rivers. The use of coal for power in the Industrial Revolution encouraged development near major coalfields. Later, connection to the national grid meant that electricity could be provided countrywide.

- **Raw materials** for traditional industry were bulky and difficult to transport (e.g. coal and iron ore). Factories and works had to be located at their source. As raw materials began to run out, many industries moved near ports where raw materials could be shipped in.

- **Suitable flat land** with space for expansion was needed for factories.

- Originally, **labour** input was low but, as processing plants and technology improved, large numbers of people migrated to fill industrial jobs.

- Factories had to locate near **markets** where goods were sold or next to major **transport routes**. Originally these were rivers and canals, then railways from the mid-1850s.

- The presence of **entrepreneurs** who had the money to set up in business but may not have wanted to move far from their homes.

- Local **innovations** influenced the development of industry in certain areas.

Industrial estates, science and business parks

In the second half of the 20th century, many industries started to relocate to large **industrial estates** on the edges of major towns and cities. Electricity was the major source of power and was supplied by the national grid. The polluting chimneys that had characterised early industrial development gave way to low-level buildings in landscaped open spaces. Inner city industries were hampered by narrow streets, and congestion increased as road transport became more important than railways. Industries located on the new estates took advantage of the new bypasses, dual carriageways and motorways being built on the rural–urban fringe.

More recent industrial developments, with different location factors, have included the creation of **business parks** – attracting companies in the finance, retail and distribution sectors – and **science parks** for biotechnology companies and those in the quaternary sector.

For these businesses, traditional factors are much less important, as they use global communication, and physical inputs are small. They may want to be near airports or the well-educated workforce of a university town. The local environment and amenities are also important in helping to attract quality staff.

DON'T FORGET

When exam questions refer to traditional industry, they mean large-scale developments starting with the Industrial Revolution. Although industrial development from 1950 may seem a long time ago, it is considered recent by examiners.

DON'T FORGET

You should have examined one detailed case study for the location of traditional industry. Examples include shipbuilding on the Clyde, coalmining in the Ruhr or iron and steel in South Wales.

Details of the location and landscape of the UK's best-known science park in Cambridge can be seen at http://www.cambridge-science-park.com/

LET'S THINK ABOUT THIS

Questions on types of industry, industrial trends or industrial systems are rare but not unknown and can be based on graphs, tables or other information. Background knowledge of these subjects can be used to enhance your answers to the more usual style of question concerning industrial location and industrial change.

INDUSTRIAL LOCATION

RECENT CHANGES IN INDUSTRIAL LOCATION

In addition to the advances in transportation and energy supply that have freed industry from certain areas, the location of raw materials has become less important over time. Heavy industries such as iron and steel have declined, and the newer factories making electronics and other goods use smaller components and raw materials that are much easier to transport.

There are other reasons for the changing face of industrial location, including:

- **Government policies** that attract industry by providing grants, rent-free accommodation and purpose-built factories.

- **European Union intervention** to encourage development in some of the poorest of the member states through financial assistance for factories and money for infrastructural development.

- The creation of **industrial estates** located near major transport routes on the outskirts of large towns and cities. Factories and businesses work together here through **economies of scale**. Many companies have relocated from cramped inner city areas to these more spacious estates.

Areas receiving special economic assistance

The UK government has, in the past, designated certain regions with names such as **Enterprise Zones** and **Special Development Areas** where extra help is given to attract investment from foreign companies. This help may include:

- subsidies to build factories or pay wages to employees

- tax breaks to reduce costs for companies for the first few years

- suspension of business rates for an agreed period of time

- training costs paid for new staff

- extra help given to companies to keep operating in times of recession.

The government may also improve regional infrastructure including ports, airports and road and rail networks.

These areas will often have high unemployment and low economic development, perhaps because of the decline of traditional industry. They are also set up to encourage industries to decentralise from their traditional areas. Often, **industrial inertia** means that companies will stay in their original location long after raw materials have run out or other factors have become irrelevant, because it is too expensive for them to move.

While the UK's position within the European Union has made it attractive to foreign companies, there is a lot of competition for investment from the rest of the continent and the rest of the world. Foreign companies that set up in the UK can just as easily move somewhere else if economic conditions in the UK become less favourable.

INDUSTRIAL LOCATION ON ORDNANCE SURVEY MAPS

Map questions often ask for the identification of certain types of industry, description of the industrial landscape or explanation of location factors, and concentrate on heavy industries and modern industrial estates.

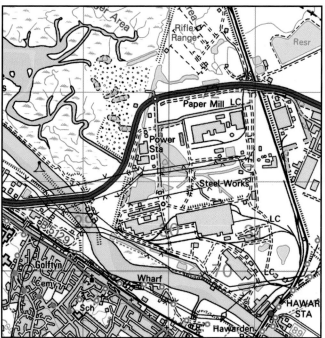

Traditional heavy industry

This site has a number of very large industrial units, including the steel works and paper mill, located on flat land near to the River Dee (which has been canalised). A main-line railway runs through the site, and there are branch lines leading to the major works. There is a power station and a number of ponds.

Location factors could include:

- Flat land that is easy to build on, with room for expansion
- A canalised river for bringing in bulky raw materials and exporting finished products to overseas markets
- A railway (and, later, the surrounding road network) to transport products to markets throughout the UK and to bring in workers from nearby towns
- Historically, a good local source of energy from the power station
- The ponds within the grounds may have provided water for cooling purposes or for use in production processes.

Heavy industries may be named on Ordnance Survey maps or referred to by terms such as **Mills** or **Wks** (meaning **works**).

Some maps have examples of industrial developments within towns and cities. These would have built up around the Central Business District and will appear cramped in a grid-iron pattern. Tenement housing may be nearby. Other identifiable features could be small power stations, gasworks, railways and derelict land.

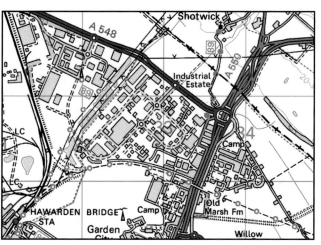

A modern industrial estate

This site has many small units and a well-planned road network with roundabouts and open spaces. It is located next to a main road (the A550) with its own junction.

On the modern industrial estate, factories and offices are on a much smaller scale employing fewer people than in traditional industries. Most people will drive to work, so roads are important, and the open spaces could indicate car parks.

This type of development is usually found on the outskirts of a town or city, and it may be possible to identify landscaping that will appeal to employees. Other location factors include nearby university towns that provide an educated workforce, and airports to bring in international executives.

LET'S THINK ABOUT THIS

The reasons why companies choose one location over another may be very complex and can change completely over time. Make sure you do not get confused between historic and recent developments and can apply your knowledge of industrial location factors to a wide variety of different businesses.

INDUSTRIAL CHANGE

THE CHANGING INDUSTRIAL LANDSCAPE

Traditional and modern industrial landscapes

Industrialisation has produced distinctive landscapes, often of large factories emitting smoke into polluted skies. In South Wales, villages grew along valley floors, merging into long linear settlements of terraced houses, collieries, slag heaps and railways.

Modern industrial landscapes are more open with car parks, gardens and water features. They are found near major transport networks and away from housing estates so that pollution is not a problem.

THE ENVIRONMENTAL CONSEQUENCES OF INDUSTRIAL ACTIVITY

- The use of coal as a power source, especially in the past, has meant that emissions of smoke, soot and dust have polluted the atmosphere.
- Waste material was often dumped in rivers in old industrial areas.
- Chemical manufacture can produce soil-contaminating pollutants.
- The waste from mining piles up in **spoil heaps** that leach toxic chemicals and are prone to landslides.
- **Deindustrialisation** leaves large areas derelict in inner city locations.
- Relocation to industrial estates on the edge of towns puts pressure on the greenbelt.

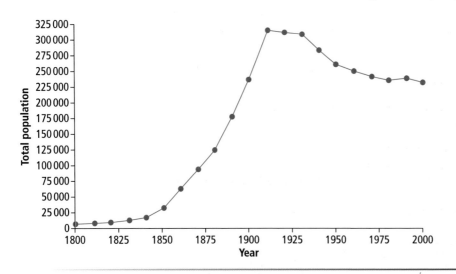

Historically, increased industrial activity led to mass migration and rapid population growth in areas like the Rhondda in South Wales as men moved from rural areas to find work in factories. These huge influxes of people also had negative impacts on the environment.

Population of the Rhondda Valley through industrial rise and decline

INDUSTRIAL DECLINE AND RISE

The developed world has seen the decline of heavy industries, known as **sunset industries**, since the middle of the 20th century. **Sunrise industries** such as electronics and telecommunications have taken their place.

Change comes for a variety of reasons. Increased competition from Asian countries meant that the shipyards of Scotland and South Wales were no longer competitive, while technological developments in the aircraft industry resulted in a decreased market for ships worldwide.

INDUSTRIAL CHANGE CASE STUDY: SOUTH WALES

South Wales is an excellent example of a traditional industrial landscape that had huge growth during the 19th century but suffered decline and industrial change from the mid-1950s. Many of the reasons for change are similar in other industrial areas.

- Rapid growth occurred because of readily obtainable supplies of raw materials, including coal, limestone and iron ore.
- New processes promoted development of the steel industry from the 1860s – this led to a growth in associated industries such as engineering and shipbuilding.
- Transport facilities could be easily established along the valley floors. Rivers were canalised and railways built to take coal and steel to the exporting ports of Barry, Cardiff and Newport.
- The expansion of the British Empire provided a huge overseas market.
- Local entrepreneurs invested heavily in industry, and there was a large labour supply to the towns and cities from the surrounding rural areas.

Causes and effects of industrial decline

From the 1950s, industry in the valleys changed.

- Raw materials began to run out. Coal mines had to be dug deeper to find the seams, and iron ore was imported – both adding hugely to production costs.
- The collapse of the British Empire led to the end of the overseas market for British steel.
- Increased competition came from cheaper products made in other countries.

As the mines and steelworks closed, job losses increased and unemployment rose. Many industrial workers moved away to find jobs in the south of England.

Crime rates rose along with unemployment and health problems, including depression, started to increase. With less economic development, services such as education and health were affected.

The landscape was filled with closed and derelict factories, mine workings and slag heaps of industrial waste, many of which leaked toxic fumes and chemicals into the atmosphere and river systems.

Industrial regeneration

Government investment since the 1970s has brought big changes to South Wales, with new factories producing electronic goods, and businesses such as call centres taking advantage of financial assistance. The relocation of government offices for the Royal Mint and the DVLA has also helped to increase employment in the area.

Other location factors such as improved transport networks (bypasses and bridges) have attracted industries looking for a prime site at the edge of Europe.

Environmental improvements have made the area more attractive for incoming workers and have provided further job opportunities:

- Old mines have been reopened as visitor centres and museums.
- Terraced housing has been renovated or replaced.
- Leisure facilities such as swimming pools, dry ski slopes and theme parks have been provided for local people and to attract tourists.
- Landscaped industrial estates and business parks have been built.

DON'T FORGET

Case studies are very important in exams – the examiners say so every year and have made particular reference to the lack of detailed knowledge in students' answers to the Industry questions in 2007 and 2008. Learn "what, when and where", to maximise your marks.

First-hand accounts and pictures of the Welsh coal mines can be found at http://www.welshcoalmines.co.uk/ and a basic interactive guide is available here: http://www.ngfl-cymru.org.uk/vtc/ngfl/ngfl-flash/steel_works_e/steel_e.html

LET'S THINK ABOUT THIS

Industry in South Wales is a popular case study, but it may not be the one you have learned in class. There will be similarities in industrial growth and decline in all case studies, but make sure you do not get them confused when answering exam questions.

SETTLEMENT SITE, SITUATION AND FUNCTION

SITE AND SITUATION

The actual point at which a **settlement** is located is called its **site**.

The **situation** of a settlement refers to its location relative to its surroundings and the factors that influenced its growth, such as routes, relief and rivers.

Historically, in western Europe, favourable sites for towns and cities included:

- Defensive sites such as the crag-and-tail of an extinct volcano, on top of a hill, in a river meander or on a headland protected by cliffs

- At a spring or near a water source on river terraces and lochsides

- On fertile land such as an alluvial fan

- At the lowest bridging point of a river or other sites where routes were concentrated – river confluence points, gaps between hills or on an isthmus

- On a dry site above surrounding marshy land

- Flat land where building could take place

- In a forest clearing where building materials were abundant

- Where harbour facilities were available in sheltered bays or river mouths

- Areas rich in raw materials such as iron ore and coal.

Although site is initially important, a settlement will not grow unless its situational factors are good. Edinburgh, for instance, held an excellent defensive site on top of a crag-and-tail but would not have grown without the surrounding features below:

- Fertile land in East Lothian that provided a major source of food

- The development of international trade through the nearby port of Leith

- The gap between the Firth of Forth and the Pentland Hills that channelled people through the city

- The early water source provided by the Nor' Loch.

The city grew quickly as wealthy Scots moved to the Royal Mile to court favour with the monarch, and businessmen and traders relocated to take advantage of the market provided by the growing population.

Edinburgh – built on a crag and tail

Shrewsbury – located in a river meander
Defensive settlement sites

DON'T FORGET

As with Industry, you need detailed case study examples for your Urban Geography. Many factors influencing settlement growth will be similar for a number of towns and cities, but having specific information will gain you marks.

SETTLEMENT FUNCTION

The **function** of a settlement is the reason it was set up and continues to exist – what it does and what activities are found within it. Almost all settlements have a number of functions (and are, therefore, **multifunctional**), although one may dominate. The bigger a settlement is, the more functions it will have and the more difficult it will be to determine which one is most important.

Settlement functions include:

- Government – especially in the case of capital cities
- Administrative – whether holding local or regional councils
- Financial – if there is a large banking or money-making sector
- Industrial – including old mining towns or places with a large number of factories
- Education – across the range from primary schools through to universities
- Medical – if the settlement contains a hospital or, at a local level, GP practice
- Entertainment – from a village hall or pub to major multiplex cinemas, museums, nightclubs and theatres
- Transport – main-line railway and coach stations may be important
- Religious – ranging from village churches to regional mosques and synagogues
- Military – for garrison towns with a high number of armed service personnel
- Resort – normally on the coast but could include spa towns, ski resorts and others
- Tourism – visitors may contribute greatly to the local economy
- Agricultural – especially for rural communities. Many market towns began by serving surrounding farms.

All settlements will also have some kind of **residential** function, often divided into low-cost housing, middle-class housing and houses for the wealthy. Rural settlements may be **dormitory towns** with a large number of **commuters**.

Most settlements, except for the smallest hamlets, will also have a **retail** function. In small villages, this may be a shop or post office, but, with cities, there will be a number of shopping centres, including large edge-of-city centres and high-class shops in the centre of town.

Case study: the changing functions of Edinburgh

Initially a **Government** and **Military** centre, Edinburgh grew in reputation as an **Education** centre after the establishment of the university in 1583. Towards the end of the 17th century, the establishment of the Bank of Scotland led to the city's growth as a **Financial** centre, and the port of Leith brought a **Trading** function with an expanding **Industrial** sector. In the latter part of the 20th century, many of these functions began to fade, but **Tourism** grew in importance – and, in the last few decades, the **Financial** sector has had a resurgence and **Retail** has boomed. Throughout, Edinburgh has maintained an important **Residential** function.

DON'T FORGET

Over time, the function of a settlement may change completely. What was once a small Cornish fishing village, for instance, may now be a famous holiday resort.

Check how Edinburgh, Glasgow and other Scottish cities have grown and changed over time with historic maps and satellite imagery at http://www.nls.uk/maps/townplans/overlays.html

LET'S THINK ABOUT THIS

The reasons why towns and cities locate in a certain area and then develop and grow are many and complex and may change completely over time. Remember to learn the factors influencing site, settlement and function so that you can interpret information you are given and apply them to your specific case study city.

URBAN LAND USE

LAND USE ZONES

There are a number of different **land use zones** that can be identified within settlements. Some of these will only have one function but others, such as the **Central Business District (CBD)**, will have several.

Factors which influence land use include:

- Age – the oldest buildings lie near the city centre.

- **Accessibility** – main transport routes meet in the CBD, but areas on the outskirts have become easier to reach as city bypasses and main roads have been built.

- **Bid-rent theory** – assumes that the most expensive land is in the CBD because it is most accessible and space is limited.

- Wealth of inhabitants – this causes distinct variations in housing quality in different zones.

- Changes in demand – once-prized industrial land near the city centre is now cramped and difficult to reach, so land use has changed.

- Planning decisions – made by the local authority.

DON'T FORGET

Being able to refer to urban models can be helpful when answering questions on the characteristics of different land use zones.

MODELS OF URBAN LAND USE

Over the years, several models have been developed to describe and explain land use zones within cities.

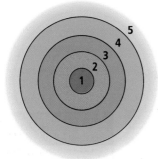

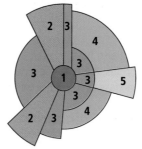

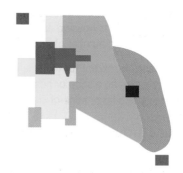

1	Central Business District (CBD)
2	Factories/Industry
3	Working-class housing
4	Middle-class housing
5	Commuter zone High-class housing

1	Central Business District (CBD)
2	Factories/Industry
3	Working-class housing
4	Middle-class housing
5	Commuter zone High-class housing

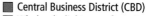

■	Central Business District (CBD)
■	Wholesale, light manufacturing
☐	Low-class residential
■	Medium-class residential
■	High-class residential
■	Heavy manufacturing
■	Outlying business district
■	Residential suburb
■	Industrial suburb

Burgess' Concentric Zone Model shows that towns grow outwards from the CBD in residential zones according to age and wealth. The model is viewed as too simplistic, as it takes no account of physical landscape and transport routes, and, when it was devised, there were no out-of-town developments such as industrial estates.

Hoyt's Sector Model improved on Burgess by showing urban zones developing along transport routes; but it is still too simple, gives zones very distinct boundaries and ignores the fact that zones may be multifunctional.

Harris and Ullman's Multiple Nuclei Model is more complex and suggests that zones grow around several different centres. Different land uses are still kept in separate zones and the CBD is still the most accessible location, but the model includes more modern land uses, such as outlying business districts and industrial suburbs.

CHARACTERISTICS OF URBAN LAND USE ZONES

The Central Business District

The CBD is traditionally the most accessible part of the city, where major road and rail networks meet. A number of functions will be found here:

- Businesses, including banks, financial services, estate agents, lawyers and solicitors

- Administration at a local, regional or national level

- Retail, with department stores, high-order shops and major fashion chains

- Cultural centres such as theatres, museums and art galleries

- Entertainment facilities, including restaurants, cinemas and nightclubs.

Tourism can also feature prominently in the CBD, and hotels are commonly found.

In many towns and cities, parks and open spaces may form part of the CBD although, with a big demand for land, this is not always the case. High rents for office and retail space means that buildings are multi-storey to provide extra space on a smaller "footprint".

The inner city

Also known as the **Zone of Transition** and the **Twilight Zone**, this is the old industrial zone of the city which grew up around the CBD. Common features include:

- Accessibility by canal, railway and, later, road networks

- Declining traditional industries such as breweries, empty mills and a number of small specialist workshops

- A range of old and new buildings, mixed land use and low-order shops

- Derelict buildings, waste land and **gap sites**

- A declining population characterised by one-person households, ethnic minorities, high unemployment and low disposable incomes.

Residential zones

These have seen many recent changes, but, originally, low-cost housing would be found within the inner city, with homes for the high number of factory workers. Terraced housing was laid out in a **grid-iron** pattern, with a high number of churches. Originally, football stadia were found in these areas (Ibrox, for instance, in Govan). In many cities, these have since moved to the outskirts.

As people became wealthier, they moved further out from the city centre to take advantage of more open space and to escape the congestion and pollution of the inner city. Housing in the middle-class zone was better quality, less densely built, with individual gardens. Street layouts were less ordered, with cul-de-sacs and curvilinear roads instead of grid-iron patterns.

Further out still were the areas of high-cost housing – very low-density detached houses and bungalows with more surrounding land. Local railways and roads gave quick and easy access to the city centre.

DON'T FORGET

Don't get confused – the inner city is not the centre of the city (that position belongs to the CBD). Zones with inner city characteristics may be found some way from the centre, e.g. Leith beside Edinburgh.

More on the characteristics of urban land use zones can be found at http:// geobytesgcse.blogspot.com/search/label/urban%20land-use

LET'S THINK ABOUT THIS

Models are useful for explaining city layout, but many have severe limitations. Being able to refer to them when discussing urban land use zones is good, but recent changes mean that there are few models that now conform to the traditional ideas.

URBAN CHANGE

URBAN CHANGE POST 1950

During the second half of the 20th century, the number of people living near city centres decreased markedly. Inner city living had become full of problems because of pollution, congestion and run-down housing.

The **migration** of the urban population is reflected in the changing characteristics of urban land use zones.

Change in the CBD

As the city centre population decreased, the permanent urban population changed to a **transient** one which would commute in, spend the day working and occasionally go out in the city at night.

The retail trade suffered and shops closed. The CBD remained the most accessible part of the city, however, so many businesses, especially chain stores, were still keen to buy premises there.

London fashion retailer Harvey Nichols has set up in Edinburgh

New pubs, restaurants, nightclubs and cinemas were set up in renovated buildings or occupied new-built premises. Covered shopping arcades (such as Edinburgh's Princes Mall) and pedestrianised areas were established, and **high-order** shops came in to serve the increasingly wealthy urban workforce.

There has recently been increased competition for businesses in the CBD from out-of-town shopping centres. These are located near major transport routes, motorways and bypasses, and are easy to reach by car. They have large, free car parks – a major advantage for consumers reluctant to pay money to park in the city centre.

The CBD has fought back, with the development of modern covered arcades providing facilities such as food courts and their own car parks (e.g. Buchanan Galleries in Glasgow). High-quality fashion shops are also being attracted into the city centres – Edinburgh, for instance, has benefited from the establishment of a world-famous Harvey Nichols store.

Change in the inner city

Factory closures in the inner city increased from the middle of the 20th century onwards. Increasing congestion had made transport very difficult, and many businesses relocated to industrial estates on the much more accessible city fringes.

Increasing unemployment as factories closed led to high rates of out-migration. **Slum clearance** was instigated because of concerns over decaying inner city housing, which lacked basic facilities such as indoor toilets. Vast numbers of people were moved to council estates and high-rise flats in other parts of the city or relocated to **New Towns**, including Cumbernauld and Livingston. Large areas of the inner city were left derelict, with ex-industrial **brownfield** sites and tenement housing in need of **renovation**.

Urban regeneration schemes have since filled gap sites with entertainment facilities (e.g. Glasgow's SECC and Science Centre) or official buildings such as conference centres. Old housing has been renovated to make luxury flats or replaced with affordable options to encourage people to move back. Recreational areas and green space for families have also been provided.

contd

URBAN CHANGE POST-1950 contd

Change in residential zones

The exodus from the CBD and inner city was accompanied by changes in residential zones further out from the centre.

- Private housing estates were built for middle-income families in the **suburbs**. These have wider streets, houses with gardens and roads set out in **cul-de-sacs** or curved patterns to provide a more pleasant environment.
- Wealthy urban residents bought large houses with gated gardens and garages in the **rural–urban fringe**. Environmental quality was high and there was good access to the city centre. These developments have courted controversy by encroaching on protected areas in the **greenbelt**.
- Large edge-of-city council estates were built with a mix of tower blocks and terraced apartments to take people cleared from the inner city slums. Many were built cheaply and soon started to decay. Few facilities or jobs were provided for residents who did not have the money to access those in the city centre. Now, unemployment rates in these estates are high, and people have been left feeling abandoned and suffering from social problems, including crime, vandalism, alcoholism and depression.

URBAN TRANSPORT SYSTEMS

Traffic problems have grown in many urban areas because of:

- increased car ownership related to higher disposable incomes
- increased commuting to city centres from the suburbs and commuter villages
- increased numbers of service buses
- increased lorry traffic to city centre shops and businesses as rail traffic has declined
- old, narrow streets and historic buildings that cannot be redeveloped.

DON'T FORGET

Many towns and cities have their own specific traffic problems and solutions – knowing a range of these will add depth to your answers.

The result has been more pollution in city centres, parking problems, congestion and time wasted in traffic jams, increased stress and occasional road rage.

Solutions developed to improve traffic management have included:

- greenways and bus lanes to encourage people onto quicker public transport
- more multi-storey car parks, use of gap sites for parking and heavy penalties for kerbside parking to prevent parked cars from further narrowing cramped city streets
- urban bypasses
- improved suburban rail services
- park-and-ride facilities on the outskirts of towns and on busy commuter routes
- development of suburban/city centre tram systems
- congestion charging and road tolls to make people pay to drive into city centres during peak times
- car-free housing developments and car-share clubs
- promotion of cycling, cycle lanes and bike-to-work schemes.

Image of what the Edinburgh trams could look like on Princes Street. Trams may be one solution to urban transport problems

Find out how London's congestion charging system works at http://www.tfl.gov. uk/roadusers/congestioncharging/ while urban transport case study material for Edinburgh and Glasgow can be found at http://www.edinburgh.gov.uk/ internet/Transport and http://www.glasgow.gov.uk/en/Residents/GettingAround/ LocalTransportStrategy/

LET'S THINK ABOUT THIS

Urban change has been driven by migration, but there are many reasons behind these population movements into and out of the city. Make sure you know how all the factors affecting urban change are related so that you can provide coherent, well-structured answers to exam questions.

URBAN LANDSCAPES

URBAN LANDSCAPES ON ORDNANCE SURVEY MAPS

Map questions can cover a variety of topics:

- Identifying the function of a settlement

- Identifying, and giving map evidence for, the CBD of a settlement

- Identifying urban land use zones and uses such as retail parks or industrial estates

- Comparing and contrasting differences in residential zones

- Explaining the factors influencing the location of all the above.

The Central Business District

Map evidence includes buildings such as town halls, museums, tourist information, cathedrals, churches and major education centres like universities.

The Central Business District of Worcester

Transport networks converge on the CBD, as it is the most accessible part of the city – so look for main roads and railway tracks together with main-line railway stations, bus and coach stations. In older settlements, the street pattern is often narrow and winding because they were built before cars were invented.

Large buildings can be identified as blocks on maps – these could indicate department stores, entertainment facilities or covered shopping centres. It is unlikely there will be any industry present in the CBD.

Residential environments

Contrasting residential environments in Cardiff

Inner city housing developments will be older, probably 19th-century, with terraced housing in a grid-iron street pattern. There will be minimal open space, railways, lots of churches and, perhaps, some old industrial development nearby. Other features may include higher-education facilities, museums, prisons and football stadia.

Modern housing estates will have cul-de-sacs and crescents and be obvious car-friendly late-20th-century developments. Houses will have gardens and there will be open spaces, perhaps woodland and parks. Roads are the most likely transport routes. There may be schools, colleges, leisure centres or other entertainment facilities.

 Logging on to the Ordnance Survey website means you can access map extracts of cities across the UK for self-practice: http://www.ordnancesurvey.co.uk/oswebsite/

URBAN CASE STUDIES

The most common urban case studies are Edinburgh and Glasgow (although London is sometimes used). Examples from the Scottish cities are included below.

The Central Business District

A large area of Glasgow's CBD, including Sauchiehall Street, has been pedestrianised for shopping, and there are covered shopping centres in the Princess Arcade and Buchanan Galleries. Edinburgh's examples include the St James Centre and Rose Street. Main-line railway stations are found in the CBD – Queen Street and Central in Glasgow, Waverley in Edinburgh. Both cities have a number of luxury hotels, civic buildings and parks in the city centre.

The inner city

Leith is the best example of change in the inner city in Edinburgh. The decreasing importance of the docks and the closure of its associated industries led to a long period of decline. Unemployment was high, residential areas were decaying and there were serious problems of drug-taking and crime.

From the early 1990s, Leith has undergone radical urban regeneration. The construction of the Scottish Office headquarters (later the Scottish Government) sparked a major boom. Old flats were renovated and new luxury apartments were built. Restaurants and hotels such as the Malmaison located in the area to serve the wealthier residents and visitors. Tourist facilities including the Royal Yacht Britannia were brought in to boost the area's image, and a major shopping centre was built at Ocean Terminal.

In Glasgow, the Clyde Gateway Project is the largest urban regeneration project in Europe and forms the centre of the 2014 Commonwealth Games village.

Edge-of-the-city

In Glasgow, slum clearance from the inner city in the 1950s and 1960s led to the growth of housing estates far away from the centre in areas like Easterhouse and Castlemilk. In these areas, thousands of people were provided with open spaces, a clean environment and new homes with indoor toilets and bathrooms. Similar schemes in Edinburgh saw the population relocated to Wester Hailes and Sighthill.

However, little thought was given to providing jobs or to the social dislocation that would result from moving people from the city centre to the outskirts. These areas became some of the most depressed in Scotland with lower-than-average life expectancy, high unemployment and a mix of other social and economic problems. Recent investment, however, has seen regeneration taking place with the growth of community projects, much-improved local amenities and transport links.

Another important development has been the growth of out-of-town shopping centres taking advantage of major transport routes. The Braehead Centre and the Fort in Glasgow and the Gyle and Fort Kinnaird in Edinburgh are all good examples.

DON'T FORGET

Many students lose marks because they answer Urban questions in general and do not give named examples of different land use zones and recent changes from their case study city.

Maps and details on the Clyde Gateway Project are at http://www.glasgow. gov.uk/en/Residents/Environment/Rivers/RiverClyde/Projects/ClydeGateway/ with the influence of the 2014 Commonwealth Games shown here: http://www. glasgowarchitecture.co.uk/commonwealth_games_village.htm

LET'S THINK ABOUT THIS

Although the general characteristics are similar across most Urban case studies, you need the detailed knowledge that applies to the one you have studied. In the exam, you will often be asked to refer to **one city** only. If you then give examples from more than one, it could cost you marks.

LANDSCAPES OF THE BRITISH ISLES

The Rural Land Resources interaction looks at some of the main landscape features of the UK, how they are exploited and the conflicts that develop between different users.

The landscapes studied are upland limestone, glaciated uplands and coastlines of erosion and deposition. Refer to the Lithosphere chapter in Physical Environments for details of the various landforms (such as corries, limestone pavement, stacks and stumps).

SOCIAL AND ECONOMIC OPPORTUNITIES IN UPLAND AREAS

Limits on economic activity

In upland areas, there are a number of restrictions on human activity – especially on ways of making money from the land. These are related to geology, climate and soils.

Glaciated uplands have steep valley sides and scree (talus) slopes that make cultivation difficult. High precipitation rates and impermeable rock result in high drainage densities, waterlogging and surface water. The cold weather restricts the growing season. Thin, infertile soils form in the harsh climate, while potentially fertile soils on the valley floors are often flooded.

Upland limestone landscapes suffer from similarly harsh climatic factors, steep slopes and hilly land. Cultivation of crops and construction of roads, buildings and houses is difficult. Waterlogging is less of a problem because of the permeability of the limestone, but the absence of surface water and the thin soils that develop hinder agriculture.

Settlement in upland areas may be limited to the flatter land found in valleys, although in many cases these are river flood plains, especially in glaciated areas. Lines of communication are also restricted because of the nature of the landscape, making upland areas remote and inaccessible.

Economic opportunities in upland areas

While there are difficulties imposed by the landscape, some types of economic activity are particularly suited to upland areas.

The steep slopes, thin soils and short growing season mean that arable farming is all but impossible except on some of the more fertile valley floors, but the pasture that grows on hillsides is capable of supporting large numbers of sheep.

Quarrying has been important in upland limestone areas for centuries. The limestone extracted is used as a building material, in road construction and in other products, including toothpaste.

Glaciated uplands provide scope for a number of other important industries:

- **Hydro-electric power** – steep valleys are easily dammed and give a good head of water for generating electricity in North Wales and the Scottish Highlands.

- Dams also hold water in **reservoirs**, acting as a resource for major population centres.

- Extensive **forestry** plantations have been planted in upland areas by the Forestry Commission since 1919. These are a major employer and income-generator in the Lake District and other areas.

- **Quarrying** of sand and gravel from kames, eskers and other fluvio-glacial deposits.

TOURISM IN SCENIC RURAL AREAS

Upland limestone and glaciated areas contain some of the most dramatic scenery in the whole of the British Isles and attract millions of domestic and foreign tourists each year. Huge numbers of visitors also head for the dramatic coastlines of Devon, Dorset, Cornwall and Pembrokeshire. Different types of tourists visit for different reasons.

contd

TOURISM IN SCENIC RURAL AREAS contd

Passive tourists drive around the countryside and view the spectacular scenery from their cars. They stop at tea shops, cafés, pubs and restaurants and visit museums (such as the Cumberland Pencil Museum in the Lake District) and other attractions like White Scar Caverns in the Yorkshire Dales. The Jurassic Coast of Dorset attracts fossil-hunters, lovers of Thomas Hardy novels and people who want to spend their days relaxing on the beautiful beaches.

More recently, the growth in **active pursuits** has seen people visiting these areas to take part in more challenging activities.

Active pursuits in selected rural areas

Activity	Location
Lochaber	Hillwalking, mountain biking, canyoning, climbing, skiing, ski mountaineering
Snowdonia	Hillwalking, mountain biking, climbing
Lake District	Hillwalking, mountain biking, climbing, watersports
Devon, Dorset, Cornwall and Pembrokeshire	Sailing, jet skiing, other watersports. long-distance coastal paths
Cairngorms	Skiing, ski mountaineering, climbing, hillwalking, sled-dog racing, mountain biking, watersports, hunting, stalking

Active pursuits in upland and coastal areas

SOCIAL AND ECONOMIC OPPORTUNITIES IN COASTAL AREAS

As well as tourism developments such as golf courses, caravan parks and campsites, and marinas for yachts, coastal areas are economically important for:

- extractive industries, including tin and coal mines and other quarrying

- power stations, increasingly using wind and tidal power

- oil terminals

- ports and ferry terminals

- military bases and firing ranges.

While bringing money into the local economy, many of these activities cause pollution and affect coastal wildlife.

 BBC's Bitesize site has some good general information and up-to-date facts on upland limestone: http://www.bbc.co.uk/scotland/learning/bitesize/higher/geography/interactions/rural_land_resources_rev1.shtml

LET'S THINK ABOUT THIS

Case study examples are all-important when looking at Rural Land Resources – common ones include the Cairngorms or Lake District for Glaciation, the Yorkshire Dales for Limestone and Dorset's Jurassic Coast. Make sure you thoroughly learn specific locations from the ones you've studied.

CONFLICT IN RURAL AREAS

Quarrying can cause conflict in upland limestone areas

Groups come into **conflict** because of the different ways in which they want to use rural areas. Limestone quarrying, for instance, has a long history and brings money into the local economy. However, the dust it produces settles on agricultural land, while blasting noise disturbs wildlife and disrupts the peace that attracts tourists. Quarry lorries block narrow country roads, cause congestion and contribute to air pollution. The quarries themselves visually pollute the landscape and are seen as **eyesores**.

In many upland areas, and in the area around Lulworth in Dorset, large areas of land are owned by the Ministry of Defence and are used for live firing. Access for the public is limited. These restrictions annoy visitors, and the noise created can scare farm animals and disturb residents. In some areas, the local environment has also been spoiled by shell fire.

The main sources of conflict in rural areas are between tourists and other user groups, even though it is often tourist money that allows them to prosper.

THE BENEFITS AND PROBLEMS OF TOURISM IN RURAL AREAS

Benefits	Problems
Tourists contribute money to the local economy	Many tourists are day trippers and spend little
Tourist money is spread through the local economy by the **multiplier effect**	
Local cafés, pubs, restaurants and hotels make money from tourists	**Honeypot** towns may develop where seasonal congestion is a problem
Gift shops and local attractions grow to cater for tourists	Gift shops may become the only viable retailers, so local people see everyday services decrease
Tourism creates jobs (in restaurants and hotels, in visitor centres and as guides, bus drivers and many other examples)	Employment may be **seasonal**, attracting migrating workers (and students) rather than local people
Local infrastructure and public transport improves to cater for tourists	More tourists visit the area, making local people feel like strangers in their own town
	Slow-driving tourists get in the way of local people, and increased traffic may cause accidents involving local people and their children
Tourists may like the area so much that they buy a **second home** and contribute more to the local economy	Second-home ownership raises house prices beyond the reach of local people
	Tourists may come into conflict with farmers because they walk over crops, disturb livestock and leave gates open, allowing animals to escape

Environmental effects of tourism

Large numbers of tourists can also damage the environment:

- In local beauty spots, visitors cause footpath erosion and may damage surrounding land (including sensitive environments such as limestone pavement) as they stray off the walkways provided.
- Tourists may drop litter in scenic areas and on farmland. This can kill wildlife or stock animals.

contd

THE BENEFITS AND PROBLEMS OF TOURISM IN RURAL AREAS contd

- Many visitors come to rural areas by car and drive slowly down the narrow country roads, causing congestion and adding to air pollution levels.
- Inconsiderate parking can damage grass verges or farmers' fields.
- Some tourist activities, such as watersports in coastal areas, create a lot of noise and can pollute valuable water resources.
- Uncontrolled wild camping may lead to fires in forested areas.

DON'T FORGET

You have to know about the features of coasts, glaciated areas and upland limestone but, when it comes to conflicts, you need knowledge of coasts and one or other of the upland landscape types.

MANAGEMENT OF PROTECTED AREAS

In order to make tourism **sustainable**, independent government-funded bodies such as **National Park Authorities** (NPAs) make rules and regulations to control activities in areas where they work. Voluntary organisations represent the views of local people and put pressure on the authorities to act.

In the Yorkshire Dales, the NPA can force quarry owners to limit their activities and order them to screen sites from public view and restore the land to a natural state when quarrying is finished. They can also encourage owners to use trains instead of lorries to transport their products.

In Dorset, Poole Council limits the environmental impacts of tourists in Poole Harbour and has set up zones for different watersports including jet skiing, fishing and yachting.

In the Cairngorms, **Scottish Natural Heritage** and the **Forestry Commission** manage large areas of land and employ rangers to educate visitors in how to minimise their impact. Litter bins have been removed to encourage people to take their rubbish home.

There are many other organisations (e.g. the National Trust) involved in the management of sensitive areas, working alongside local landowners such as the Rothiemurchus estate near Aviemore.

DON'T FORGET

There will be certain environmental problems specific to the case studies you have looked at – those caused by the expansion of the skiing industry and the building of the funicular railway in the Cairngorms National Park are good examples.

Examples of protected land in the UK

ESAs	Environmentally Sensitive Areas	Areas of agricultural land where farmers are encouraged to "safeguard and enhance country with a high landscape, wildlife or historic value"
SSSIs	Sites of Special Scientific Interest	The country's very best wildlife and geological sites – over half of them are internationally important in conserving habitats
AONB	Areas of National Beauty	"A precious landscape whose distinctive character and natural beauty are so outstanding that it is in the nation's interest to safeguard them"

Sources: www.naturalengland.org.uk, www.sssi.naturalengland.org.uk, www.aonb.org.uk

The impact of European Union and UK government policies

Some environmentally sensitive areas attract grants and subsidies from the government and the EU. Farmers are paid to maintain their land in traditional ways, promote conservation and look after features such as drystone walls. They are also given money to diversify their businesses and provide more facilities for tourists. Fertiliser and pesticide use may be limited in return for cash payments, and land can be **set aside** to provide habitats for wildlife.

Details of the UK's National Parks and what they do to manage the impact of tourism can be found at http://www.nationalparks.gov.uk/ and Dorset's Jurassic Coast at http://www.jurassiccoast.com/

 LET'S THINK ABOUT THIS

If the exam question is asking about conflicts, make sure you mention both sides of the argument. If it asks about problems, then only one side is required. Read the question carefully before answering.

RURAL LAND DEGRADATION

Degradation occurs when soils are so affected by physical and human factors that they cannot support either their natural vegetation or the farming of crops and animals. This deterioration in productivity has serious social, economic and environmental consequences.

The Rural Land Degradation interaction uses three main case studies: North America is compulsory and is studied with either Africa north of the Equator or the Amazon basin.

It ties in with Physical Environments through the climatic effects of the ITCZ (Atmosphere) and the study of soils (Biosphere). It also connects with the Rural topic of Human Environments by studying the effects of shifting cultivation in the Amazon and extensive commercial agriculture on the plains of North America.

SOIL EROSION

Population growth, especially in the tropics and sub-tropics, is putting huge pressure on rural resources. There are more people who need to be fed, and more farmers in need of land to grow food and feed them. Soils are being put under intense pressure, and farmers are moving on to more **marginal** land on the edges of their traditional agricultural regions. **Climatic variability** means that soils in these areas are quickly farmed to produce crops if climatic conditions (particularly rainfall) are favourable, but they lose fertility and degrade when they are not.

In semi-arid areas of savanna, grassland and bush, where degradation has been particularly severe, desert-like conditions (**desertification**) have spread. This is often exacerbated by climate change, especially in sub-Saharan Africa.

The main factors influencing how quickly a soil will erode are:

- **Vegetation** – plant roots bind the soil and protect it from wind and rain, so the more dense the plant cover, the less erosion will take place.
- **Relief** – steep slopes are eroded much more easily than gentle ones.
- **Soil type and soil structure** – soils with large sandy particles have much higher infiltration rates and are less susceptible to erosion by water. When they are dry, fine soil particles such as clay and silt are more easily blown around by the wind.

Inappropriate agricultural practices also cause soil erosion. **Overgrazing** – where too many animals all feed in the same area – destroys the vegetation cover and soil structure. Growing the same crop over large areas (**monoculture**) or growing too many crops year after year (**overcultivation**) strips nutrients from the soil and renders it incapable of supporting plants. Fertilisers and pesticides contaminate the soil and kill important soil organisms.

When the plant cover is removed and soil structure is damaged, wind and rain are the main factors responsible for the actual erosion of the soil.

Gully formation can permanently change the landscape

contd

SOIL EROSION contd

Processes of soil erosion by water

- Raindrops hitting the soil surface dislodge particles and displace them around the point of impact or move them downslope. This **rainsplash** effect increases with the heavier drops of tropical thunderstorms.
- Rainsplash can result in the air spaces in the soil becoming clogged with smaller particles. Infiltration rates decrease, and water flows over the surface as **sheet wash** (overland flow) instead, dislodging more soil particles.
- Water flowing downhill also enlarges and erodes natural channels into temporary **rills** and permanent **gullies** – these can be several metres deep and wide.

Rain hitting the soil and water flowing as sheet wash **detaches** soil particles, **transports** them downhill and **deposits** them at the foot of a slope or in a river or lake. The same three-step process is important in wind erosion.

Processes of soil erosion by wind

If vegetation cover is removed and the topsoil is dry (as happens in arid and semi-arid areas which experience dry seasons), erosion by the wind occurs. It is particularly important in areas of flat land, where the size of fields is large and where there are strong winds such as the Harmattan that blows south of the Sahara.

Dust storms carry soil in suspension from Africa

The biggest soil particles may be forced to slide or roll along the ground as **surface creep** if the wind speed is particularly strong. Smaller particles can be bounced or jump along the ground by **saltation**. Soil particles jumping through saltation may dislodge other particles when they land.

The most fertile and finest particles of the topsoil can be blown thousands of miles (even from Africa to southern England) in **suspension**. This can be seen on satellite pictures of upper-atmosphere dust storms. Particles will only be deposited when the wind speed drops or they are washed out of the atmosphere by rain.

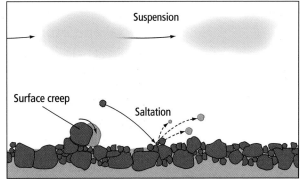

The processes of soil erosion by wind

DON'T FORGET

Soil carried in suspension can be a very dramatic sight and brings to mind pictures of Dust Bowl conditions in 1930s USA. However, it is only responsible for 15 per cent of **aeolian** erosion. Saltation, accounting for 70 per cent, is much more damaging.

DON'T FORGET

A basic knowledge of erosion processes is important in Rural Land Degradation, but you need to be able to apply this knowledge to the specific case studies asked for in the exam questions.

Facts and figures on the global problem of desertification can be found at http://www.ifad.org/events/wssd/gef/gef_ifad.htm#t5

LET'S THINK ABOUT THIS

Rural Land Degradation questions often have a resource you have to study, describe and interpret. For example, you may be given a graph showing changing rainfall in the Sahel and be asked how this might lead to land degradation.

CASE STUDIES OF RURAL LAND DEGRADATION 1

The processes of soil erosion and land degradation are similar in all three case study areas, but the causes, social, economic and environmental impacts and management strategies are quite specific to the different regions.

LAND DEGRADATION ON THE GREAT PLAINS OF NORTH AMERICA

Causes

Overcultivation of the North American prairies resulted in disastrous land degradation and the **Dust Bowl** conditions of the 1930s:

- Demand for wheat from both within the USA and overseas led to an increase in farming of marginal areas.
- Natural prairies in the west were overgrazed down to their roots as cattle herd sizes increased with demand for beef.
- **Monoculture** of wheat stripped the land of nutrients and destroyed soil structure.
- Inappropriate farming methods such as burning stubble and ploughing at the wrong time of year left soils open to erosion.
- Overploughing created **furrows**, which encouraged erosion by water.
- A succession of unusually dry years led to the dusty soils being susceptible to processes of aeolian erosion.

Erosion was so severe because the land was flat and the lack of trees gave little protection for the soil. In many areas, rainfall was less than 400 mm per year, leaving the soils very dry, loosely bound and easily blown away.

Since the 1930s, farming methods have improved greatly, but the soil structure is still being destroyed by the overuse of fertilisers and pesticides.

Social and economic impacts

- Loss of livelihood for farmers
- Knock-on effects of reduced business for related services, including shopkeepers and grain merchants
- Widespread migration and abandonment of the prairie lands
- Rural depopulation resulting in the closure of schools and hospitals
- Health effects such as increased depression and suicide rates
- Countrywide economic impacts through lost income from foreign exports, reduced tax revenue and the need to provide financial support to those affected.

Environmental impacts

The loss of millions of tonnes of topsoil during the Dust Bowl years was the most dramatic environmental effect. Dust clouds blew over long distances, settled on vegetation, buried homes and blocked out the sun.

Increased use of chemicals since the 1930s has contaminated underground water supplies, soils and rivers and affected the natural vegetation and wildlife.

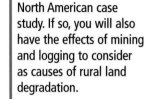

DON'T FORGET

You may have used the Tennessee Valley as your North American case study. If so, you will also have the effects of mining and logging to consider as causes of rural land degradation.

Dust Bowl conditions on the North American prairies

Plenty of film clips of the Dust Bowl years can be found at http://www.youtube.com/watch?v=x2CiDaUYr90

LAND DEGRADATION IN AFRICA NORTH OF THE EQUATOR

Causes

Population increase in the African **Sahel** has caused huge pressure on the land. With more people:

- The need for **fuelwood** increases. Trees are already a meagre resource, and deforestation leaves the dry soil vulnerable to wind and water erosion. Urbanisation has also increased the need for fuelwood in towns and cities.
- There are more animals. **Overgrazing** removes the plant cover, while animals trample and compact the earth, especially around water holes, leading to greater surface runoff.
- There is an increase in farming of **marginal** areas on the edge of the Sahara. Although this can be fine in years of above-average rainfall, periods of **climatic variability** and failing rains lead to rapid degradation.
- **Overcultivation** is common, as there is a need to produce more food. This leads to a reduction in fallow periods, destroying soil fertility.

Map showing the location of the Sahelian countries

In some countries, governments have encouraged farmers to turn to monoculture and **cash crops** to increase export revenue. Inappropriate farming methods, including **flood irrigation**, are used and strip the soil of nutrients.

When soils are degraded and exposed, they dry out quickly and are easily eroded by the wind. Rain falls as intense **convectional** downpours which can wash away the topsoil, even on quite gentle slopes, and leads to rill and gully formation.

Social and economic impacts

Malnutrition, starvation, increased **infant mortality** and widespread **famine** can occur if crops fail year on year. Other impacts include:

- loss of income and declining living standards for farmers
- failing rural education systems, as farmers cannot afford to send their children to school
- increasing cost of grain and **staple food** crops
- more reliance on **foreign aid**
- loss of traditions as **nomadic pastoralists** are forced to settle down
- conflict between different groups as pressure on the land increases
- migration, especially of young men, leading to a rural **gender imbalance**.

Environmental impacts

- Farming in marginal areas combined with increasing drought frequency has resulted in the southward migration of the Sahara. Over the last 50 years, 65 million hectares of the Sahel have turned to desert.
- There is an increase in rills and gullies across the region.
- Topsoil is lost, especially when blown by the dry Harmattan wind. This may be deposited on crops in more fertile areas.
- **Salinisation** has increased in areas where irrigation has been used to grow cash crops.

> **DON'T FORGET**
>
> The case study of Africa really only refers to those countries immediately south of the Sahara. Learn their names and some specific examples of the impacts of land degradation in individual countries.

A case study of land degradation around Lake Chad can be found at
http://edcwww.cr.usgs.gov/earthshots/slow/LakeChad/LakeChad

LET'S THINK ABOUT THIS

The study of rural land degradation has taken on increased relevance with changes in the global climate and as more and more people have been affected – it has even been implicated in armed conflict such as that in the Darfur area of Sudan.

CASE STUDIES OF RURAL LAND DEGRADATION 2

LAND DEGRADATION IN THE AMAZON BASIN

DON'T FORGET

While Brazil's exploitation of the Amazon brings such huge economic benefits, deforestation will be very difficult to control.

Causes

Traditional farming methods used by shifting cultivators have used the rainforest sustainably, but recent migration and exploitation from a number of activities have vastly increased **deforestation**.

- South America's increased population and wealth has led to a demand for beef catered for by the large **cattle ranches** spreading across the region.
- **Mining** of tin, gold, silver and other minerals under the Amazonian soil generates wealth. The search for oil is under way in the upper courses of the Amazon basin.
- Tropical **hardwoods** are taken from the forest and, for every one removed, up to 30 others are knocked down as it is dragged from the forest.
- Successive Brazilian governments have encouraged **settlers** to take up **peasant farming** in the rainforest. These settlers quickly degrade the soil and move to a new plot to start the process again.

Mining is a big cause of soil degradation in the Amazon

Precipitation in the rainforest is at least 2500 mm per year (almost four times that in Edinburgh), and much of this falls as heavy tropical downpours. When the vegetation cover is removed, trees no longer intercept the rain, and topsoil is washed away. Intense leaching of nutrients further degrades the soil.

Social and economic impacts

- Shifting cultivators are forced to move as pressure on the land increases.
- Shifting cultivators may turn to **sedentary** agriculture. **Fallow** periods are lost, and crop yields fall as the land loses its fertility.
- **Indigenous communities** may no longer be able to feed themselves.
- Tribes lose their identity and traditional customs, migrate to cities and add to the problems caused by rapid urbanisation.

Environmental impacts

The loss of nutrient-rich topsoil is just one of the environmental problems caused by deforestation:

- Increased runoff increases sediment levels and **silting** in rivers and affects levels of flooding.
- Wildlife habitats and important plant and animal species are lost.
- Increased global warming may result from forest burning and the loss of the Amazonian **carbon sink**.

SOIL-CONSERVATION METHODS IN THE CASE STUDY AREAS

Many solutions have been proposed to stop soil erosion. To be **effective**, these need to be **appropriate** for the area where they are practised – methods which involve a large capital outlay will not work well in poor Sahelian countries.

Suitable methods for Africa

- **Afforestation** programmes to reduce wind erosion, protect the soil from erosion by water and provide fruit, nuts and fuelwood
- **Bench terracing**, **contour ploughing** and the construction of **stone lines** following the contours of the land all reduce runoff
- **Gabions** to check gully erosion
- **Biogas** stoves to reduce reliance on fuelwood
- Construction of **stone dams** to trap water in seasonal river beds
- Large- and small-scale irrigation projects

Effective solutions must be low-cost, sustainable and acceptable to the whole community.

Suitable methods for the Amazon basin

- **Reafforestation** and the protection of existing forests by conservation groups, both national and international, and the setting up of National Parks
- Increased global cooperation to help Brazil protect the Amazon
- Sedentary farmers using **crop rotation** to maintain soil fertility
- Agroforestry methods to encourage reforestation of degraded land

The pressures on the Amazonian rainforest are immense because of the wealth the region can generate. For soil-conservation methods to be effective, cooperation will be needed between many countries so that financial alternatives to its exploitation can be given.

Suitable methods for North America

- Contour ploughing and **laser levelling** flatten the land and reduce the amount of soil removed by overland flow
- **Minimum tillage** (where the use of machinery is limited) reduces damage to the soil
- **Shelter belts** – where trees are planted to help stop wind erosion
- **Intercropping**, **strip cultivation** and fallow periods to maintain soil fertility
- **Cell grazing**, with the land providing animal food rotated around a central point, reduces the damage caused by overgrazing
- Overgrazing can also be prevented by ranching bison instead of cattle
- **Centre-pivot irrigation** to stop drought conditions developing
- Education and improved farming practices

Many methods of soil conservation in North America are very expensive, and government subsidies are important in making them effective.

Building stone lines is a low-cost community-based solution

DON'T FORGET

It is very rare to be asked about soil-conservation methods without having to evaluate their effectiveness and relevance to the case study area where they are used.

DON'T FORGET

Rural Land Degradation questions follow a fairly standard format asking about causes, effects and solutions in the case study areas – familiarise yourself with what's expected by looking at past papers.

 Appropriate community solutions to land degradation in the Sahel can be found at http://www.excellentdevelopment.com/

LET'S THINK ABOUT THIS

Solutions to land degradation differ greatly depending on the case study area, but there are some common themes. Education in appropriate farming techniques and restriction of agriculture in marginal areas are low-cost ways of reducing soil erosion. Irrigation can be an effective solution but has negative effects because it increases salinity and reduces the water held in underground **aquifers**.

RIVER BASIN MANAGEMENT

Human impacts on drainage basins take place on a range of scales and have profound effects on the hydrological system. The consequences of large-scale **water-control projects** such as **dam-building** can be both positive and negative.

The River Basin Management interaction requires detailed knowledge of a case study from North America, Africa or Asia. The most common river basins studied are the USA's Colorado River and the Nile in Egypt.

There are many links with the Hydrosphere unit in Physical Environments.

DON'T FORGET

If questioned on the distribution and characteristics of major river basins, include description of the number of large rivers and the directions they flow in as well as explanation of the general patterns.

DISTRIBUTION OF RIVER BASINS AND WATER-CONTROL PROJECTS

The sources of major rivers are usually found in mountain ranges such as the Rocky Mountains or the Ethiopian Highlands, where precipitation rates are high.

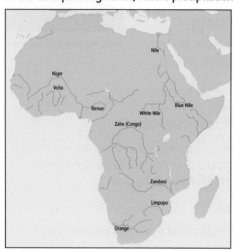

The major river basins of North America and Africa

Whether or not a river needs management through water-control projects depends on a number of factors:

- the number of people living downstream needing a regular supply of drinking water
- the livelihoods of people in the catchment area
- major industrial or agricultural developments that depend on a regular water supply
- weather patterns and climatic variability in the drainage basin.

CONTROL OF FLOW IN DRAINAGE BASINS

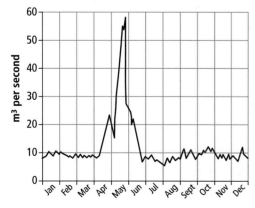

The natural annual discharge of the Colorado river

Many rivers have variable discharge levels throughout the year and are affected by spring snowmelt or seasonally intense heavy rain, especially in arid and semi-arid areas. This gives a number of problems for people living in the drainage basin:

- **Seasonal droughts** affect agricultural productivity and drinking water supplies.
- High discharge levels over a short period of time may cause flooding.
- Transport may be affected if the river is a major **navigation** route.

Controlling the flow of a river (for instance, by building a dam) can reduce the peaks and troughs in discharge, giving an even and reliable flow throughout the year. This maintains irrigation for agriculture and reduces the risk of flooding. Damming a river provides a source of **hydro-electric power**, and the reservoir formed behind the dam may be used to develop recreational facilities.

INTERPRETING HYDROLOGICAL DATA

To assess the need for river basin management, it may be necessary to interpret a variety of maps, graphs and diagrams from a number of different sources.

Discharge hydrographs

Historical records of discharge can show the effects of dam-building on the peaks and troughs in river flow. Over a shorter time scale, they show important information such as lag time and the slopes of the rising and falling limbs, which may give clues to the drainage basin characteristics.

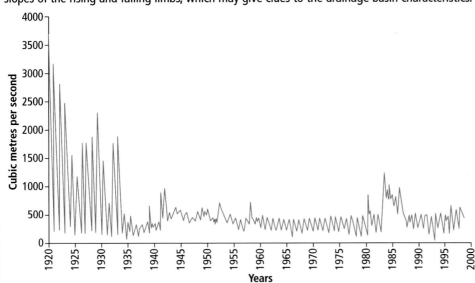

Annual discharge of the Colorado river downstream of the Hoover Dam (completed in 1935)

Climatic data

Climate graphs show seasonal changes in rainfall and the times of the year when low rainfall may pose problems. Graphs from different parts of a river's course help identify areas suffering from a **water deficit** that may need irrigation transfer systems from further upstream. Evaporation rates are also important – in arid and semi-arid areas, high evaporation rates may mean that low precipitation levels are ineffective.

Drainage basin maps

Maps can show population centres susceptible to flooding and agricultural areas that need a regular irrigation supply. The location of dams and water-control centres can also be seen. A map displaying the underlying geology will show the areas of impermeable rock most suitable for dam construction.

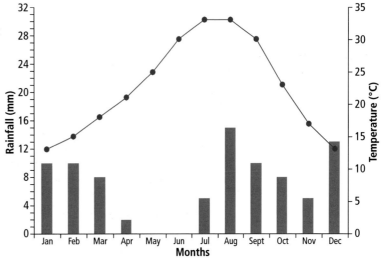

Climate graph for Yuma, Arizona

 A global introduction to river basin management can be found at http://www.slideshare.net/aland/river-basin-management-part-1

LET'S THINK ABOUT THIS

To describe the distribution of river basins on a continental scale, you will need a good basic geographical knowledge of countries (or states in the USA) and places where rivers enter oceans, seas or lakes. Get an atlas out and start learning your stuff.

WATER CONTROL PROJECTS

DAM-BUILDING

Both physical and human factors have to be considered when selecting a site to build a dam.

Physical factors

- The underlying rock must have a solid, stable **foundation** and should be relatively impermeable to prevent leakage of water.

- The site should be unaffected by earthquake activity and **subsidence** to avoid the risk of damage to the dam.

- A narrow canyon, gorge or valley will keep the length of the dam as short as possible but should be deep enough to contain a sufficient volume of water.

- Low evaporation rates will help maintain water levels.

- A high enough water flow from the upstream catchment is needed to replenish supplies.

- Accessibility is required during the dam construction phase and for subsequent maintenance.

Human factors to be considered include:

- The number of settlements that would be flooded during construction and the cost of relocating their inhabitants

- The historical or cultural sites that would be flooded

- The amount of farmland that would be lost through flooding and the compensation costs payable to farmers

- The distance from end users such as consumers of hydro-electric power (HEP), farmers needing irrigation water and towns and cities looking for a regular drinking supply.

THE EFFECTS OF DAM-BUILDING

Dam-building affects the hydrological cycle of a river basin and has positive and negative social, economic and environmental impacts.

Effects on the hydrological cycle

- Reduced water flow behind the dam

- Increased evaporation from the surface of reservoirs

- Infiltration rates affected by water held in the reservoirs and irrigation channels

- Rivers may be diverted or canalised

- **Microclimatic** changes in the local area

- Fluctuating water tables

- Changes in the seasonal variations of river levels.

Social effects

Irrigation and a regular water supply may lead to an increase in agricultural productivity, a better diet, improved health and less disease among the local population. HEP can give a more efficient, cleaner and more environmentally friendly power supply than was previously available. Recreational opportunities may increase, from watersports on the new reservoirs to walks in the surrounding area.

> **DON'T FORGET**
>
> The exact benefits and adverse effects of water control projects will depend on your case study – learn it well and don't deal in generalisations if asked for specific examples.

contd

THE EFFECTS OF DAM-BUILDING contd

Negative effects can include the forced removal of people from their homes in the areas to be flooded and an increased incidence of water-related diseases such as schistosomiasis and malaria in the reservoir and connected irrigation channels.

Economic effects

Improved farming can lead to an increase in local sales and export revenues. Industry may benefit from a local HEP supply, increasing local employment and money gained from taxes. A more constant flow of water upstream may make navigation and river transport easier and more reliable.

The main negative effect is the huge cost of building a dam. Increased taxes may be needed to pay for such schemes in developed countries, while developing countries may have to rely on foreign aid and long-term debt. Irrigated agricultural land may need more fertiliser, increasing the costs for farmers. Vital communication links in rural areas can be disrupted or destroyed by dam construction.

Environmental effects

- Fresh water supplies will improve sanitation and health among local people.

- Flood control may save thousands of lives and precious agricultural land, but valuable river **alluvium** no longer adds to soil fertility in annual floods.

- In some areas, large reservoirs are seen as a scenic improvement, while in others they are judged as eyesores.

- Increased industry and agriculture may pollute water courses.

- In areas where rivers contain large amounts of sediment, reservoirs and dams become clogged with sediment, rendering them ineffective or incurring heavy maintenance costs.

DON'T FORGET

Some effects, such as increased malaria, are only relevant to developing countries.

DON'T FORGET

Many effects of dam construction are connected, so the silting of dam machinery has an economic cost and the provision of recreational facilities has an economic benefit – mention these connections in your answers.

POLITICAL CONSEQUENCES OF WATER-CONTROL PROJECTS

If river basins cross political or state boundaries, there is the potential for conflict:

- There may be a reduction in flow for areas downstream of dams.

- Water pollution may cross political boundaries.

- One area may gain control over another's water supply.

- Countries or states may have to share the cost of dam construction.

Complex legislation and international or interstate agreements may be needed to ensure fair distribution of the benefits of water-control projects. In the USA, the **Colorado River Compact** addresses this issue, and, in East Africa, the **Nile Basin Initiative** has been signed by ten countries with a stake in the river to ensure that all have a fair allocation of water.

A detailed look at the effects of dam-building can be found at http://www.ehponline.org/qa/105-10focus/focus.html

LET'S THINK ABOUT THIS

Dam-building can be a controversial global business. Your River Basin Management system case study could involve forced relocation of farmers in China, disputes over water rights between Iraq, Syria and Turkey or the abuse of British aid money to finance controversial dam projects in Malaysia.

RIVER BASIN MANAGEMENT CASE STUDIES

The two most popular case studies are the River Nile in Africa and the Colorado River in the USA. The basin characteristics are very different, but there are some similarities in the impacts that water-control projects have had on the rivers.

CASE STUDY 1: THE COLORADO RIVER

DON'T FORGET

To fully appreciate the effects of water control projects, you need to have a detailed understanding of the hydrological cycle (inputs, transfers and outputs) and how it relates to your case study river.

Basin characteristics

The source of the river is in the Rocky Mountains, where precipitation can be up to 2500 mm per year. Summer temperatures are high and lead to a net loss of water through evaporation. Peak discharge levels are influenced by snowmelt in April and May. Historically, before water control projects were introduced, floods were common and are still possible when rainfall is influenced by extreme events such as **El Niño**. The river flows through deep sandstone gorges and canyons which make ideal storage sites for dams.

The drainage basin of the Colorado River

Over the last 50 years, there has been huge population growth and urbanisation in the south-western USA, largely due to the availability of water from the Colorado River. There has been a dramatic increase in the water supply demand of urban areas and irrigation water needed for the cropped lands of California and cattle ranching in the upstream states.

Water control projects

The first agreement on building dams to ensure a steady flow of water was made almost 100 years ago (in 1922). The 11 major dams in the seven states the river crosses (and Mexico) are necessary for:

- flood control
- hydro-electric power to fuel industrial expansion and increased employment
- irrigation for farmers in Utah, Wyoming, Arizona, Nevada, California and Mexico – a total of 800 000 hectares of land is irrigated under the Colorado management scheme
- urban water supply – 70 per cent of San Diego's urban water, for instance, comes from the Colorado
- recreation – Lake Powell in Glen Canyon is the largest water recreation area in the south-west USA.

The effects of dam-building

Water control on the Colorado River has meant:

- Flooding is now absent or vastly reduced on valuable agricultural land south of the Hoover Dam
- A relatively even annual discharge profile
- The destruction of wilderness areas, bringing opposition from the strong US environmental lobby
- Threats to traditional Native American sites such as Rainbow Arch in Utah
- Increased deposition of sediment in lakes, affecting the ecology of the region
- High **salinity** in irrigated water supplied to farmers in Mexico
- An almost complete lack of wildlife in the Colorado delta, once home to jaguar, marine species including porpoise and important bird species.

contd

CASE STUDY 1: THE COLORADO RIVER contd

Future concerns for the management of the river include the need for more control over supply to urban areas, the cost of water to farmers, whether there should be greater protection for the environment and cultural heritage, and who will take charge of the flow.

CASE STUDY 2: THE RIVER NILE

Basin characteristics

The River Nile has two sources. The **Blue Nile** flows from Bahr Dar in the Ethiopian Highlands where rainfall is seasonal, reaching a peak of over 400 mm in July but with a distinct dry season from December through to February. The Blue Nile is joined by the **White Nile** in Sudan. This branch flows northwards from Jinja in Uganda where rainfall is more evenly spread throughout the year, with a peak of just under 300 mm in April. The different rainfall regimes ensure that the Nile flows all year round.

Historically, the Nile flooded annually, and this was responsible for maintaining the fertility of the ancient agricultural lands of Egypt. Along its course, population densities are much higher in the areas where its water is used for irrigation than in the desert found on either side.

The River Nile's Aswan Dam

The effects of building the Aswan Dam

The most significant water control project on the Nile is the **Aswan Dam**, originally built in 1902, with the latest version finished in 1970. The dam has meant:

- Increased irrigation for Egyptian agriculture, and multiple crops each year
- The creation of hydro-electric plants that initially provided over half of Egypt's electricity
- Flood mitigation in the 1960s and protection from drought in the 1970s and 1980s
- The development of a fishing industry in Lake Nasser – now struggling because of distance from markets
- A lack of alluvial river silt being deposited on agricultural land to the north of the dam and an increased reliance on artificial fertiliser instead

- Silting of Lake Nasser, reducing its capacity to hold water
- Increasing salinity of irrigation water and chemical pollution due to the reliance on fertilisers to maintain soil fertility
- The flooding of priceless archaeological treasures in southern Egypt
- The flooding of Nubia and the destruction of the lifestyles of its nomadic pastoralists and their historic agricultural lands
- The loss of Egypt's red-brick construction industry which once used the replenished sand of the Nile delta.

Additional effects are being felt in the Mediterranean, where a lack of sand and silt washed into the sea is causing coastal erosion and a decline in fish stocks.

Many links to river basin management projects worldwide are listed on http://www.sagt.org.uk/54

LET'S THINK ABOUT THIS

Unlike most of the Environmental Interactions, you only have to learn one case study for River Basin Management. This means you must have a high level of in-depth knowledge, as you may have to refer to it for every part of a 50-mark question. There is only very general coverage here – you will need much more detail to answer questions on the social, economic and environmental benefits of management projects and their adverse consequences.

URBAN ENVIRONMENTS

In 2007, for the first time in human history, there were more people living in towns and cities than in rural areas. The urban environment is in constant change, and the increase in population brings huge pressures, especially in the developing world.

Two case study cities are needed for the Urban Change interaction – one from the developed world and one from the developing world. A thorough knowledge of the Urban section of Human Environments is needed, together with relevant information from the Industry and Population topics.

DISTRIBUTION OF URBAN CONCENTRATIONS

DON'T FORGET

You could be asked to describe the distribution of towns and cities for a specific country. Using your knowledge of site and situation, you should be able to do this for anywhere in the world.

Cities have generally developed:

- near coasts or major lakes and rivers
- where routes are concentrated
- next to raw materials
- on flat land
- in areas with a favourable climate.

They are less likely to be found in rocky landscapes, at high altitudes and in areas affected by extreme weather.

In some parts of the world, specific factors are important. Across Western Europe, people moved to take part in industrial development on the coalfields. In Brazil, settlement is sparse in the inhospitable Amazon rainforest.

AN INCREASINGLY URBANISED WORLD

Before 1950, almost all the world's most populated cities were in the developed world, but more recently there have been huge increases in **urbanisation** in developing countries.

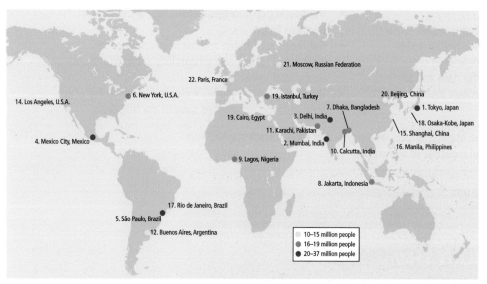

The world's biggest cities in 2015

Cities such as London and Glasgow grew because they were trading centres and attracted industrial development. In the 1800s and early 1900s, people flocked to these urban centres for the jobs they offered. Birth rates increased, while medical advances and improved sanitation cut infant mortality and extended life expectancy. New York and other cities in the New World were also immigration entry points.

Urbanisation in today's developing world has increased for similar reasons. Cities including São Paulo, Lagos and many more in India and China are experiencing high birth rates and decreasing death rates, increasing employment from urban industries and immigration from rural areas.

There are also more specific reasons for the differences in growth patterns between the developed and developing worlds.

contd

AN INCREASINGLY URBANISED WORLD contd

Reasons for urban growth in developing countries

- Cities attract large numbers of migrants because they are perceived as having better education, health and employment opportunities than rural areas – the "bright lights" effect.
- Rural populations are more affected by war, famine, drought and competition for agricultural land – push factors that force emigration to urban areas.
- Urban migrants are normally younger, and the fertility of this **youthful population** leads to high birth rates.
- Improved diet, clean water and better health care leads to decreasing infant mortality and lower death rates in developing world cities.
- Vaccination programmes and other medical advances are more readily available in urban areas than in the less accessible countryside.

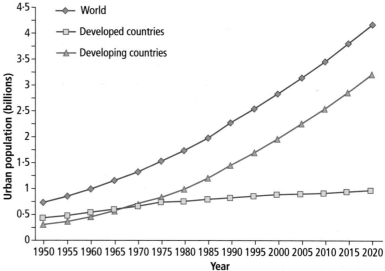

Increasing urbanisation since 1950

Reasons for urban stagnation in developed countries

Most cities in the developed world now have very low growth rates – in some, such as Berlin, they are starting to decline.

- Birth rates are decreasing, especially among urban professionals who are leaving marriage and child-rearing until later in life to concentrate on their careers.
- Increasing car ownership and improved public transport means that people can commute to work from rural towns and villages some distance away.
- People no longer have to live in cities because of the wider dispersal of jobs to science parks, business parks and edge-of-city retail centres.
- Planning legislation often limits the outward growth of cities.

> **DON'T FORGET**
>
> Cities in developing countries are growing because of both immigration and natural population increase.

Counter-urbanisation

In some countries in the developed world, there has been much more recent movement from urban areas to the countryside. Families especially are looking for a better quality of life, selling expensive properties in the cities and buying large houses with gardens.

Over the last 40 years, large numbers of people have also been given the opportunity to leave the decaying inner city and move to jobs and houses in **New Towns** (e.g. Livingston or East Kilbride).

Both these movements have led to a decreased urban population and the growth of settlements in rural areas. This outward migration is known as **counter-urbanisation**. Recently, efforts have been made to reverse this trend by improving facilities in urban areas, renovating inner city housing and redeveloping areas by building luxury flats and apartments. This repopulation of the inner urban areas is called **gentrification**.

 Video clips of New Towns and other aspects of urban change can be viewed at http://www.bbc.co.uk/scotland/history/scotlandonfilm/media_clips/index_topic. shtml?topic=newlife&subtopic=new_towns

LET'S THINK ABOUT THIS

Many aspects of the Urban Change interaction have already been described in the Human Environments topics – site and settlement, urban demographics and inner city decay. Learn how to apply this knowledge to the different kinds of question you will have to answer in the second exam paper.

URBAN CHANGE IN DEVELOPED COUNTRIES

SHOPPING

The urban shopping **hierarchy** consists of:

- The CBD, where land values are highest, **accessibility** is greatest and the most important shops are found – department stores and high-end fashion retailers (e.g. Harvey Nichols and Louis Vuitton in Edinburgh)

- Tourist shops near to major attractions – the Royal Mile running from Edinburgh Castle, for instance, has many kilt and tartan businesses

- Secondary shopping centres containing supermarkets and lower-end fashion stores located in residential areas throughout the city

- Neighbourhood shopping centres found as **linear developments** along busy roads with chemists, hairdressers and smaller supermarkets such as the Co-op and Tesco Metro in areas with a large population or high pedestrian footfall

- Isolated general stores on housing estates.

The biggest growth in retail over the last two decades has been **out-of-town shopping centres**, locating off bypasses and motorways on the outskirts of cities. They have a huge range of stores, with a few big names such as Ikea or Marks and Spencer, and may have food courts and entertainment facilities to encourage longer visits from customers. Examples include the Gyle and Kinnaird Park in Edinburgh and the Fort and Braehead in Glasgow.

These centres need lots of land for buildings and car parks, are very accessible and are often found near a large potential workforce and customer base.

Glasgow's Fort is one of Scotland's newest out-of-town shopping centres

 There is plenty of information on out-of-town shopping centres on the internet. Try Glasgow Fort: http://www.glasgowfort.com/ or the Gyle: http://www.gyleshopping.co.uk/index.shtml

THE CHANGING CBD

Competition from out-of-town shopping centres, together with the decreasing population caused by counter-urbanisation, has led to the decline of shopping in the CBD. Traffic congestion and high city centre parking fees have also had an effect.

Edinburgh City Council has tried to stop this decline by increasing bus services to make travel to the CBD easier and by offering incentives to luxury retailers to relocate there.

THE CHANGING INNER CITY

Industry abandoned the inner city from the mid-20th century onwards because of transport congestion, lack of room for expansion and, in the case of the dockyards of London, Liverpool and Glasgow, increasingly cheap competition from overseas.

The **gap sites** left have recently been redeveloped into office blocks, luxury apartments, affordable housing schemes, entertainment and leisure facilities. The relocation of the BBC headquarters to Glasgow's old dockyards and of the Scottish Government to Leith has improved employment opportunities and the image of these areas.

contd

THE CHANGING INNER CITY contd

Derelict inner city housing has been demolished and the inhabitants often relocated to edge-of-city housing estates such as Edinburgh's Wester Hailes and the Glasgow schemes at Castlemilk and Drumchapel. Local people moved to these areas often felt **dislocated**, and this has led to a lack of **social cohesion**. The inner city areas they left behind were redeveloped to provide better housing for the young professionals attracted to work in the new jobs that were created.

DON'T FORGET

There is more detail on characteristics and change in urban land use zones and on transport problems and solutions in the Urban topic of Human Environments.

EDGE-OF-CITY DEVELOPMENTS

City bypasses and other improvements in transport infrastructure have made the outskirts of the city more accessible and have led to a number of developments:

- Out-of-town shopping centres.

- **Business parks** attracting companies with large numbers of office staff to areas such as Glasgow Business Park near Baillieston.

- **Science parks** – Heriot-Watt University established the first science park near Edinburgh in 1971. A more recent development has been the growth of a biotechnology centre at Little France on the other side of the city.

- Large retail warehouses for supermarket chains such as Lidl and food companies including Fyffe's have located alongside the M8 near Livingston.

- **Multinational companies** set up sites for their own headquarters in attractive suburban locations. The RBS world headquarters at Gogarburn has a Starbucks, Tesco Metro, a florist, chemist, crèche, gym and fitness centre and landscaped gardens to try to attract the best workers.

Royal Bank of Scotland World Headquarters at Gogarburn, near Edinburgh

Conflicts on the rural–urban fringe

Despite legislation introduced in the 1950s, the areas of **greenbelt** (land on the rural–urban fringe which is protected from development) are under pressure from **urban sprawl**. Acceptable development of the greenbelt can include recreational and educational facilities, but, increasingly, economic development is being allowed, causing the following problems:

- Increased demand for housing has led to encroachment on the greenbelt, property speculation and compulsory purchase of land.

- Increased commuting has brought traffic congestion, pollution and road-building on formerly protected land.

- Pressure on agricultural land and vandalism by residents of peripheral housing estates has caused problems for farmers.

Urban residents complain about the loss of recreational land as the greenbelt is developed. Environmentalists and conservationists campaign heavily against the loss of greenbelt land and say that redevelopment of inner city **brownfield** sites would be better.

Often, development leapfrogs the greenbelt and encroaches on rural settlements further out. This leads to conflict with residents who feel the character of their towns and villages is changed by the incomers and have to deal with problems such as increased traffic. Local people may no longer be able to afford the inflated house prices, and facilities including schools may be unable to cope with the increased population.

DON'T FORGET

There has to be a balance between preservation and economic development – without the expansion of Edinburgh Airport on greenbelt land, for instance, the economy of the whole of Scotland could suffer.

LET'S THINK ABOUT THIS

When outlining the solutions to urban problems such as traffic, changes in the CBD and inner city and pressure on the rural–urban fringe, you may also need to comment on how effective they have been, referring to specific named examples – again, look back to the information in the Urban topic of Human Environments.

URBAN CHANGE IN DEVELOPING COUNTRIES

URBAN LAND USE

Models of urban land use for developing world cities look similar to those for the developed world, but there are marked differences. Identified zones include:

- the CBD
- high-cost housing, comprising old colonial villas and luxury apartments near the centre and extending outwards along major communication lines with high-security gated communities, flats and detached houses
- the **periphery**, consisting of lower-class housing
- the **shanty towns** (**favelas** or **squatter settlements**) – poor quality, **informal** housing, self-built on the poorest quality land
- factory zones along transport routes and on the outskirts of the city may be surrounded by shanty towns housing workers.

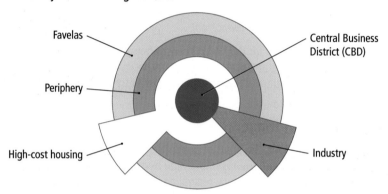

Model of urban land use in a developing country

CHARACTERISTICS OF DEVELOPING WORLD CITIES

The current growth rate of developing world cities is much quicker than that previously experienced in the developed world. Natural population increase and migration from rural areas are causing social, economic and environmental problems.

- People are attracted by perceptions of a better quality of life, but the reality is often very different.
- Employment opportunities are limited – many cities, such as São Paulo, have a booming industrial sector but the jobs on offer are far fewer than the number of migrants looking for work.
- Many people end up working in the **informal** sector (e.g. selling goods to car drivers, fixing and mending and shoe-shining) with little financial reward.
- Schools, hospitals, water and power supplies and other urban infrastructure cannot cope with the population growth.
- Increased numbers of private vehicles and poor quality public transport cause gridlock and pollution problems.
- There are huge variations in housing quality, from luxurious penthouse flats to poorly constructed shacks – often right next to each other.
- The disparity between rich and poor can lead to jealousy and an increase in crime, civil unrest and rioting.

The United Nations, the World Bank and foreign aid programmes have put a lot of money into fighting urban poverty, but many cities, especially those in Africa and the poorest Asian countries, show little sign of improvement.

SHANTY TOWNS

New arrivals from rural areas often have little money and are forced to build their own homes using whatever materials they can find – corrugated iron, plywood, cloth and cardboard. These squatter settlements quickly grow but lack facilities such as electricity, water and refuse collections. Open sewers run through the streets, and diseases such as **cholera**, **malaria** and **typhoid** can reach epidemic levels. Other problems include:

DON'T FORGET

Shanty towns have different names in different countries – **favelas** in Brazil, **barrios** in Mexico and **katchi-abadis** in Pakistan.

- Landslides, floods and pollution because shanty towns are built on the cheapest land – unstable hillsides, flood plains or rubbish tips which are unusable for anything else
- High population densities, with whole families occupying one room
- Poor communications, with few bus services to jobs in the distant city centre
- High unemployment
- The possibility of being bulldozed by city authorities cracking down on illegal land occupation
- Soaring crime rates, including drug trafficking, as shanty towns become no-go areas for the police.

Kibera shanty town in Kenya

Makeshift settlements house millions of people worldwide. In Karachi (Pakistan), there are more than 500 holding over 7 million people, while Kibera (Kenya), the largest slum in Africa, is home to more than a million.

Despite all the problems, shanty towns are crucial in providing homes to rural migrants and people on low incomes. Over time, a strong **community spirit** often develops and the quality of life improves.

Solutions to shanty town problems

As settlements become more established and residents find regular jobs, they make improvements to their homes using better quality materials such as bricks. Electricity generators and water tanks may be installed, and community **self-help** groups carry out construction projects to install sewerage systems and organise refuse collections. Other help can come from outside agencies:

DON'T FORGET

Your case study city will have its own urban problems and shanty towns – you will be expected to refer to it in-depth and not just deal in general information. Popular case study examples include São Paulo, Nairobi and Karachi.

- Established shanty towns may get recognition from the government or local council, who then install basic facilities and improve roads and street lighting.

- The World Bank and other agencies may fund self-help house building schemes.

- Some authorities have tried relocating settlements and have even offered financial incentives to residents willing to move.

Self-help schemes tend to be successful, as they build on local community spirit; but relocation schemes often fail, as people prefer to stay put in the areas they know rather than move to isolated settlements erected by the authorities.

Long-term, the best solution to shanty town problems is to decrease immigration by improving the situation of poverty-stricken rural areas. This is very difficult for poor developing countries to achieve.

 An excellent exercise on the development of a shanty town is available at http://www.sln.org.uk/geography/geoweb/blowmedown/shanty05.swf

LET'S THINK ABOUT THIS

Questions on Urban Change cover a range of topics, but you will have covered many of them as part of the Urban unit. Revise that topic at the same time as this one, and get your case study information right – it is always identified as a weak point by examiners.

INEQUALITY IN THE EUROPEAN UNION

There are physical, social and economic reasons why some European Union countries are richer than others and why inequalities exist within countries. A variety of policies have been devised to reduce these inequalities and the problems they cause.

The European Regional Inequalities interaction has some connection with the Population and Industry units of Human Environments but mainly stands on its own as a new course of study. Country case studies must include the United Kingdom and one other European Union **member state**.

PATTERNS OF REGIONAL INEQUALITY

Austria	Latvia
Belgium	Lithuania
Bulgaria	Luxembourg
Cyprus	Malta
Czech Republic	Netherlands
Denmark	Poland
Estonia	Portugal
Finland	Romania
France	Slovakia
Germany	Slovenia
Greece	Spain
Hungary	Sweden
Ireland	United Kingdom
Italy	

European Union member states

Historically, the countries of the **European Union (EU)** have been divided into a richer **core** and the poorer **periphery** countries, with **intermediate** regions lying between them.

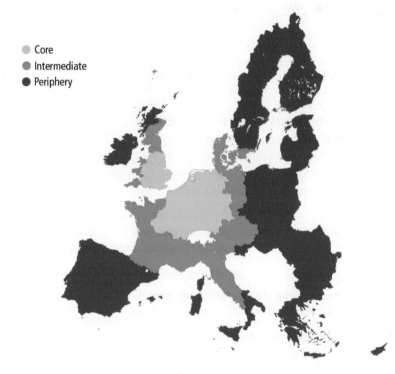

- Core
- Intermediate
- Periphery

Core, intermediate regions and periphery of the European Union

The core is the area with **high economic potential**. It covers the most densely populated part of Europe and has the main concentration of manufacturing industries. The major markets are located here together with the most highly developed infrastructure. Originally, this area grew rich because of access to raw materials, including coal and iron ore.

The peripheral areas have low economic potential, are less well developed than the core and, being more remote from markets, have a lower population density. A greater proportion of the population is involved in agriculture, and countries often have difficult physical and climatic conditions. Most of the newer, Eastern European members of the EU are peripheral countries. Intermediate areas, meanwhile, have moderate economic potential.

Recently, the distinction between the core and periphery has become increasingly blurred. The economic performance of core countries such as Germany and Italy has decreased while that of some of the peripheral countries, including new entrants such as Estonia, has grown.

SOCIO-ECONOMIC INDICATORS OF INEQUALITY

A number of different social and economic **indicators** are used to highlight the differences between regions. These include:

- the percentage of the population who are unemployed
- the percentage of the population employed in primary, secondary, tertiary and quaternary industries – the **employment structure**
- the financial support given by the EU per head of population
- **Gross Domestic Product** (**GDP**), which measures how much money a country makes in a year per person (**per capita**)
- measures of the cost of living and quality of life, such as house prices and car ownership
- **economic growth rate** (% per year), measuring how fast a country's economy is growing.

Indicators are useful, but it is difficult to judge development by using only one of them. High unemployment may be a result of increasing industrial mechanisation. This could lead to a high economic growth rate and is not necessarily a sign of underdevelopment. A low GDP may be less relevant in countries where the cost of living is low and there are government-sponsored benefits such as free health care and education. Analysis of a number of different indicators is needed to understand the economic development of a country.

FACTORS LEADING TO PATTERNS OF INEQUALITY

Physical factors

- **Climate** influences agriculture, determines areas of drought or abundant water and influences the development of a tourist industry – Spain, a country with many semi-arid areas, is a prime example.
- **Relief** is important, as mountains make it difficult to build roads and buildings, while flatter areas, although prone to flood, are generally more fertile.
- **Geology** influences surface drainage, soil formation and agriculture and determines the types of raw materials, mining and quarrying available. The availability of coal was, historically, a major development factor in countries including the UK and Germany.
- **Location** is important, especially with increasing distance from markets.

Human factors – social

- Areas of high **population density** make better markets for goods and services – many of the highest densities lie within countries in the EU core.
- **Industrial decline**, leading to unemployment and poverty, makes an area unattractive to relocating companies.
- **Isolation**, whether perceived or real, discourages business and industry from locating in an area and affects some countries on the EU periphery.
- **Levels of education** and **skills** influence the quality of the available workforce.

Human factors – economic and political

- **Infrastructure** (roads, railways, ports etc.), often related to government investment, can make an area more accessible and decrease isolation.
- Levels of **investment**, whether from individuals, companies, governments or the EU, influence the range of industry in an area.
- A balanced employment structure is important, as areas with high agricultural employment but little industry tend to be poorer.
- Support from governments and the EU and the designation of special **economic zones** increase outside investment.

Facts, figures and maps showing EU regional inequalities can be seen here:
http://personal.lse.ac.uk/RODRIGU1/Cambridge%204.pdf

LET'S THINK ABOUT THIS

Questions often involve analysis of multiple sources of socio-economic data showing the patterns of regional inequalities across the EU. You will need to use all the information to compare and contrast regions and verify whether statements made are accurate.

CASE STUDIES OF REGIONAL INEQUALITIES

Regional inequality exists on a national as well as an international scale.

REGIONAL INEQUALITY IN THE UNITED KINGDOM

UK average	£165,412
Scotland	£134,191
Northern Ireland	£149,769
The North	£123,526
The North-West	£127,929
Yorkshire and the Humber	£123,100
Wales	£135,155
West Midlands	£156,211
East Midlands	£140,671
East Anglia	£171,506
The South-West	£176,894
London	£253,751
The South-East	£214,480

Average house prices in the UK

In the UK, there is social and economic disparity between the richer south and the poorer north. This **north–south** divide is identified from a number of indicators, with GDP per capita, average house prices, employment rates and car ownership all generally higher in the south. There are some anomalies in this pattern – London, for instance, has a higher unemployment rate than Scotland and some other northern regions.

People living in the south are likely to be better educated, have more access to doctors and dentists and have better general health. Rates of obesity and coronary heart disease are likely to be higher in parts of the north.

Reasons for the north–south divide

- The north has more mountains, poorer soils, more rainfall and lower temperatures than the flatter, climatically milder south, where fertile soils have encouraged greater agricultural development.
- The north is more remote, less accessible and further from the major markets of Europe.
- Broadband internet, mobile phone reception and other business connections are less reliable in the far north.
- Natural resources, including coal and iron ore, have either run out or become too expensive to exploit, leading to a decline in traditional northern industries such as shipbuilding.
- The northern workforce has more of a reputation for strikes and industrial action, making it less attractive for industry.
- Higher population densities in the south mean bigger markets for goods and services.
- Proximity to the Euro-core has encouraged investment and the development of new **sunrise** industries such as biotechnology and electronics in the south of the country.
- The infrastructure in the south is better developed, with more international airports, ferry terminals, motorways and the Channel Tunnel providing European connections.

Problems caused by the UK's regional inequalities

Higher unemployment in the north has led to poverty, deprivation, increased crime and drug use and declining health. Services, including shops and entertainment facilities, have closed down because of a lack of money – this leads to a further increase in unemployment. Derelict factories and ex-industrial sites are eyesores and may pollute the atmosphere, surrounding land and local water courses.

> **DON'T FORGET**
>
> Your case study of a European Union country will have similar characteristics, causes and problems to those in the UK – but make sure you know the specific details, facts and figures.

DIFFICULTIES FACED BY PERIPHERAL REGIONS

Problems such as low investment in industry, poor infrastructure, long-term poverty, high crime rates, high unemployment and low wages are similar to those identified for the UK's north–south divide. Other specific difficulties in the Euro-periphery include:

- a high percentage of the workforce employed in primary industry, especially agriculture
- increasing emigration of the **economically active** workforce, leading to an **ageing population**
- dependency on welfare payments and aid from the EU
- deprivation and increasing numbers of children suffering from malnutrition and living in poverty.

SOLUTIONS TO REGIONAL INEQUALITIES

Governments and the EU have devised solutions to tackle regional inequalities, although there may be clashes between these national and international policies.

National-level solutions

Many governments combat inequalities in their own countries by encouraging investment in deprived areas. The UK government has:

- set up **special development areas**, **enterprise zones** or **assisted areas** where relocating companies (including multinationals) receive benefits such as grants, rent-free factories or free power and reduced taxes for a set period of time

- set up agencies such as **Scottish Enterprise** to combat economic deprivation

- encouraged improvements in infrastructure by building bridges and new roads or investing in regional airports and ferry terminals, such as Rosyth in Fife, to improve connections with Europe

- provided specialist help and assistance to key industries, especially manufacturing

- given training grants to the long-term unemployed and money to help them set up their own businesses

- relocated government departments to areas with high levels of unemployment.

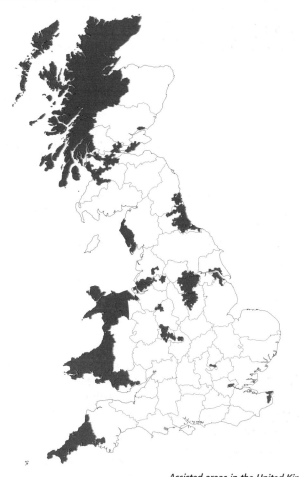

Assisted areas in the United Kingdom 2007 to 2013

European Union solutions

The EU provides large sums of money to help areas of high deprivation or isolation. The current budget of around 50 billion euros a year is distributed from three main sources:

- The **European Regional Development Fund** gives money for infrastructural projects to the poorest regions (e.g. the Eden Project in Cornwall).

- The **European Social Fund** helps with job creation and training, especially for under-represented groups, including women in Spain and the Roma in Romania.

- The **Cohesion Fund** is reserved for infrastructural projects aimed at improving transport, renewable energy and the environment (by funding, for instance, the European Rail Traffic Management System).

Most of the money is given to countries with a GDP less than 75 per cent of the EU average, although all countries qualify for help for their poorest regions. Historically, other money was available through the **European Investment Bank** and EU **Objective One** funding for remote rural areas, old industrial areas, long-term unemployment and youth unemployment.

DON'T FORGET

You should be able to quote statistics showing whether or not these solutions have brought about any noticeable improvements in order to comment on their effectiveness.

 More information on the EU's policies to fight regional inequality can be found at http://europa.eu/pol/reg/overview_en.htm

---

LET'S THINK ABOUT THIS

There have been big changes in the European Union since 2004, when many Eastern European countries joined. You need to make sure you have up-to-date examples of problems and solutions from these newer member states in your answers.

DEVELOPMENT AND HEALTH

DON'T FORGET

You need to be able to explain why certain indicators do little to explain the quality of life in a country – so make sure you are fully aware of what each indicator means and how it applies in specific countries.

There are physical, social and economic reasons why some countries are more developed than others and why inequalities exist within countries. Levels of health, **adult literacy** and average income per person are all good indicators of the level of development.

The Development and Health interaction uses information from the Population unit of Human Environments, including the influences on life expectancy, birth, death and infant mortality rates. Case study examples are needed from a variety of developing countries, and there is an in-depth look at one major disease – **malaria, bilharzia (schistosomiasis)** or **cholera**.

MEASURING DEVELOPMENT

A range of **development indicators** is used to measure the levels of development between and within countries.

Economic indicators

- **Gross Domestic Product (GDP) per capita** – the total value of all goods and services produced within the borders of a country, divided by the number of people. Similar indicators include **Gross National Product (GNP)** (GDP plus the money gained by residents of a country from investments overseas) and **Gross National Income (GNI)** (which reflects the average income per person in a country)
- **Employment structure** – the percentage of people employed in primary, secondary, tertiary and quaternary industry
- Percentage of people employed in agriculture
- Value of **exports** and **trade balance** (**surplus** or **deficit**)
- Consumption of electricity.

Social indicators

- Indicators of health, including **infant mortality rates** (the number of children dying under one year of age per 1000 live births), **life expectancy** (in years), **number of people per doctor, access to safe water** and **average calorific intake**
- Indicators of education, such as **percentage adult literacy** rates, **access to higher education** or **percentage of pupils completing primary education**
- Population indicators, including **birth rates** (per 1000 total population), **death rates** (per 1000 total population), **levels of urbanisation, population structure** and **population growth rates** (as a percentage)
- Lifestyle indicators, such as mobile-phone use and access to the internet
- Measures of political freedom, such as **freedom of speech** and **access to justice**.

Composite indicators

Traditionally, a country's development was judged by economic indicators; but this is now thought of as too simplistic. Oman, for instance, has a high GDP because of its oil reserves, but education and literacy rates are low. Cuba, meanwhile, is poor but has good health and education services.

Composite indicators use a combination of indicators to give a better idea of development. Originally, the **Physical Quality of Life Index (PQLI)** was the main composite indicator, but the United Nations now uses the **Human Development Index (HDI)**. This ranks countries by combining:

- Gross Domestic Product per capita – adjusted to take account of the different cost of living in different countries
- Life expectancy at birth
- Adult literacy rate and levels of enrolment in primary, secondary and tertiary education.

All the countries of the world are then placed on a scale from the most to the least developed.

Country	HDI rank	Adjusted GDP per capita (US$)	School enrolment index (%)	Life expectancy (years)	Adult literacy (%)
Iceland	1	36,510	95·4	81·5	100
United States	12	41,890	93·3	77·9	100
United Kingdom	16	33,238	93	79·0	100
Cuba	51	6000	87·6	77·7	99·8
Oman	58	15,602	67·1	75·0	81·4
Brazil	70	8401	87·5	71·7	88·6
Equatorial Guinea	127	7874	58·1	50·4	87
Zimbabwe	151	2038	52·4	40·9	89·4
Malawi	164	667	63·1	46·3	64·1
Sierra Leone	177	806	44·6	41·8	34·8

Development indicators for selected countries

DIFFERENCES IN DEVELOPMENT

There are common factors influencing the levels of development in different countries.

- **Climate** – areas with extreme weather are likely to be less developed and may suffer from drought and famine.
- **Landscape** – countries with forbidding terrain, for instance mountains or rainforests, are likely to have isolated populations.
- High **population growth** means that more money has to be spent on education, health and food for the young, non-productive population.
- Prevalence of **disease** can decrease worker productivity and direct national budgets towards health care rather than economic development.
- **Trade** – landlocked countries will have fewer opportunities for trade.

- **Industrial development** – manufacturing industries tend to increase wealth, while high agricultural employment often signifies low levels of development.
- **Political corruption** hampers development, while political stability encourages investment.
- **Natural resources** often increase a country's wealth.
- **Natural disasters** such as hurricanes and earthquakes affect many poor countries, which find it difficult to rebuild houses and public services.
- **War** and **conflict** can direct resources away from public services.
- **Historical influences** – countries with ties to colonial powers may have better-developed trade links.

DON'T FORGET

Define your indicators exactly – birth rate is not enough, the correct indicator is birth rate *per 1000 total population*.

Many of these factors affect how much money a government will have to invest in its industry, infrastructure and public services; but, even if a country has positive characteristics, it may not be developed. Some countries have large reserves of natural resources, but these may be difficult to exploit or be owned by foreign companies, so bring little wealth. In other countries, such as Equatorial Guinea, political corruption means that the money gained from natural resources is taken by the few people in power while the majority of the population lives in poverty.

DON'T FORGET

If asked to explain differences in development, examples are important, but don't just reel off a list of countries with one reason why each is rich or poor – you need to avoid generalisations and show real understanding of why a country is at a particular stage of development.

NEWLY INDUSTRIALISED COUNTRIES

Countries where the contribution of industry to GDP exceeds one third of the total are called **Newly Industrialised Countries (NICs)**. These are countries that have achieved a very high rate of economic growth over the last 30 years – in many cases outperforming developed countries in terms of manufacturing output. Brazil and Mexico were among the first NICs, but most are found in south-east Asia, including countries such as Malaysia and Thailand. Many of these countries saw the success of Japan following the Second World War and wanted to improve their own living standards. Their governments achieved this by long-term industrial planning, concentrating on:

- adding value to exports by processing primary products
- developing a manufacturing industry – originally concentrating on heavy industries such as shipbuilding then on micro-electronics and high technology
- encouraging local entrepreneurship and backing this up by attracting investment from foreign multinationals

- increasing regional trade by forming the Association of South-East Asian Nations (ASEAN)
- looking after the main natural asset – a resourceful and relatively cheap workforce – and investing in its education.

The huge growth led to the original Asian NICs (Hong Kong, South Korea, Taiwan and Singapore) being dubbed the **tiger economies**. They have since been joined by others including Malaysia, Thailand, the Philippines and Indonesia and the **emerging economies** of India and China. In all these countries, average incomes have increased, but some have retained a number of developing country characteristics, including significant numbers of people living in poverty.

More information on development indicators and the Human Development Index can be found at http://hdr.undp.org/en/statistics/

LET'S THINK ABOUT THIS

Questions on development indicators are very common, so you need to know how useful individual indicators are at showing levels of development within a country and be able to describe the precise indicators used to make up a composite index.

DIFFERENCES IN HEALTH AND DEVELOPMENT

DIFFERENCES IN DEVELOPMENT WITHIN COUNTRIES

Development indicators give average figures for the whole of a country, but there may be huge differences between regions within a country. These occur at several levels:

- between isolated and less remote rural areas
- between rural and urban areas
- between the urban rich and the urban poor.

These differences are often more exaggerated in large countries such as Brazil.

Many developing countries span different climatic zones, so areas with favourable weather for agricultural development will be richer than those with the extremes experienced in high mountains or rainforest. Distance from markets and major population centres will also impact on rural poverty. In Kenya, the rural areas of the central highlands grow cash crops such as tea, fruit and flowers, while **subsistence farming** or **nomadic pastoralism** is dominant in the semi-arid areas to the north.

Urban areas have more industrial development than rural areas and are generally wealthier. There is better access to education and health facilities in towns and cities, and infrastructure tends to be of a higher quality. Water and power supplies are also much more reliable in towns and cities. Cash-strapped governments can also tend to neglect rural areas in developing countries because the literate electorate lives in the urban areas.

Contrasts exist within urban areas. Many developing world cities have modern office blocks and luxury apartments housing the wealthier occupants. Often, these are located next to shanty towns with high levels of poverty and unemployment, low levels of education, high crime rates, poor health facilities, poor sanitation and epidemic diseases.

DON'T FORGET

Read the question carefully, and don't mix up differences in development **between** and **within** countries – you could be asked about either.

DON'T FORGET

You will need a detailed case study example showing the regional differences in development within a country – common examples are Brazil and Kenya.

FACTORS INFLUENCING THE SPREAD OF DISEASE

Physical and human factors determine both the types of diseases and their prevalence in specific areas.

- Climate is the most important physical influence on the incidence of disease. Hot, wet conditions encourage disease development and provide ideal breeding conditions for those spread by **insect vectors**.

- The lack of a clean, safe water supply encourages disease development. In rural areas of some developing countries, communities face a long walk to collect water from a stagnant pond. Urban areas may have more regular supplies, but access is often denied to the poorest sectors of society.

- In shanty towns where basic sanitation is absent, disease is spread by contaminated sewage running through the streets. In such conditions, **infectious** diseases can quickly spread.

- Access to health care and disease prevention is difficult in isolated communities in mountainous areas, rainforests and deserts.

The most important influence is poverty. Poor people tend to live in overcrowded conditions where disease spread is rapid. They have low levels of education, so know little about disease prevention. **Immunisation programmes** and access to medicines is also lacking in poor developing world countries.

contd

FACTORS INFLUENCING THE SPREAD OF DISEASE contd

Poverty leads to a lack of vitamins, carbohydrates or proteins in the diet and **malnutrition**. People suffering from malnutrition are more susceptible to disease and find it harder to fight off illness. Once ill, they are unable to work and make money and become even poorer. Ill-health, disease and malnutrition particularly affect the young and the old, increasing infant mortality rates and decreasing life expectancy.

```
              Malnutrition
Low agricultural output          Increased susceptibility
and food shortages                    to disease

Little investment in                    Decreased ability
agriculture and                            to work
agricultural industry

Less tax take and          Poverty      Unemployment and
decreased development                    a lack of money
```

The spiral of poverty and disease

PRIMARY HEALTH CARE

Developing countries cannot afford expensive medicines and hospitals for their citizens, so rely on cheaper methods of **primary health care**, including:

- advice on **diet** and **food preparation** and the provision of common medicines such as **anti-malarials** and **oral rehydration therapy** in small local clinics

- 'barefoot doctors' or **medical auxiliaries** – rural residents trained in basic medical knowledge who can treat straightforward illnesses and refer more serious cases to doctors and hospitals

- large-scale **vaccination programmes** (e.g. for measles) initiated by rural medical auxiliaries

- **family-planning** and **child health-care clinics**, including **pre-natal** and **ante-natal care** for mothers and children in rural towns and larger villages

- the development of local **health education programmes** and small clinics in shanty towns in urban areas

- the use of cheap local alternatives and traditional medicines instead of expensive ones manufactured overseas

- aid programmes by organisations such as **Christian Aid** concentrating on improving the general health of the urban and rural poor.

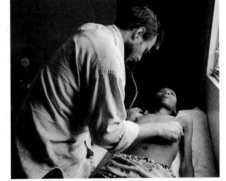

A rural health clinic in Malawi

Some countries have set up networks of rural health centres that are part-funded by local communities. In Benin, vaccination programmes for BCG, polio, diphtheria and tetanus carried out in these rural clinics have been effective for large parts of the rural population. Costs for the local community are limited to half a dollar per visit.

DON'T FORGET

You need case studies of primary health care strategies for a number of different countries and an idea of how effective they have been. Examples include barefoot doctors in China, health education programmes in Nicaragua and Christian Aid's urban programmes in India.

Examples from Oxfam of primary health care: http://www.oxfam.org/en/content/primary-health-care-services-poor-countries-photo-gallery

LET'S THINK ABOUT THIS

Lack of clean water and inadequate sanitation are major factors in spreading disease – but don't forget the full range of physical and human influences that lead to variations in infant mortality, malnutrition and life expectancy.

DISEASES OF THE DEVELOPING WORLD

WATER-BASED KILLERS

Poverty and harsh climates lead to a lack of clean drinking water supplies, exacerbated by poor sanitation and sewerage. These factors are responsible for many of the most common diseases in the developing world. These can be characterised as:

- **water-borne** – when people using contaminated water catch and spread diseases, including cholera
- **water-based** – when parasites use hosts living in the water to transmit disease, e.g. bilharzia
- **water-related** – when insect vectors such as mosquitoes breed in water then transmit diseases such as malaria.

Many of the diseases spread through contaminated water cause **diarrhoea**. According to the **World Health Organisation (WHO)**, diarrhoea is responsible for 4 per cent of all deaths worldwide and is a major killer of children because of **rapid fluid loss**. It increases infant mortality rates and lowers overall life expectancy in many developing countries.

DISEASE CASE STUDY 1: MALARIA

Malaria infects over half a billion people each year and causes more than one million deaths, more than 90 per cent of them in Africa. Despite this, it is treatable and preventable.

Distribution

Malaria is common in the tropics and sub-tropics including India, south-east Asia, tropical Africa and areas of South and Central America.

Infection and symptoms

The female anopheles mosquito

The disease is caused by **parasites** called **plasmodium**, carried by the **female anopheles mosquito**. The mosquito bites an infected human (the **host**), passing parasites into the blood which then multiply in the liver. After two weeks, the parasites leave the liver and attack red blood cells, causing: headaches, nausea and stomach pains; fever; kidney failure; swollen spleen; lethargy; convulsions; coma and, if untreated, death.

Factors encouraging malaria

Certain **environmental conditions** are needed for the female anopheles mosquito to breed:

- Air temperatures between **15°C** and **40°C**
- Areas of **shade** for the mosquito to **digest** its blood meal
- The mosquito lays its **larvae** on **stagnant water surfaces**, including lakes, ponds, padi fields, irrigation channels, puddles and wells. Standing water increases in areas of high rainfall.
- A **human blood reservoir** – densely populated areas are ideal.

Human factors can increase the likelihood of infection:

- Lack of investment in sanitation, health care and education increase the risk of disease spread.
- The amount of stagnant water is increased by human activities – creation of reservoirs, irrigated farming and, after wars and conflicts, unfilled **bomb holes** and **craters**.
- Shanty towns are often built on poor-quality **marshland**.

Using bed nets to fight the spread of malaria

contd

DISEASE CASE STUDY 1: MALARIA contd

Controlling malaria

Outside Africa, malaria was almost wiped out in the 1960s and 1970s by the worldwide WHO attack on the disease. Mosquito resting areas were sprayed with **DDT**, and hosts were treated with **anti-malarial drugs** such as **chloroquine**. DDT use was abandoned when it was found to damage humans and the environment, and the parasite developed **immunity** to drugs, so malaria started to increase again. A number of different approaches are now used with varying success:

- Health and education campaigns in rural villages
- Supplying insecticide-sprayed **bed nets** to human blood reservoir populations
- New anti-malarial drugs such as **atovaquone** and those developed from compounds found in traditional remedies such as Chinese **artemisinin**
- **Genetic engineering** to release **sterile** male mosquitoes
- **Drainage** of breeding sites
- Dropping **mustard seeds** in padi fields to drown the mosquito larvae
- Adding fish to padi fields to eat the larvae
- Spraying ponds with **egg white** to suffocate larvae
- Using new insecticides, e.g. **Malathion**

DON'T FORGET

As well as knowing these methods of control, you need to have an idea of how effective they are. Genetic engineering is very expensive and unlikely to be a success, but supplying bed nets is a cheap option that could be easily carried out.

DISEASE CASE STUDY 2: CHOLERA

Cholera bacteria are spread through water contaminated by human waste. They are mainly found in India, south-east Asia and, more recently, South America and Zimbabwe. Infection rates are uncertain, but up to half a million people per year may develop mild or moderate symptoms. **Severe watery diarrhoea** and **dehydration** can occur, and this occasionally causes **fatalities**.

Cholera is always a risk in areas of severe poverty, or after wars and natural disasters when people are crowded together in refugee camps. It is treatable with **oral rehydration therapy**. The best methods of control are the provision of clean, safe **chlorinated** water supplies and effective sanitation systems. Most importantly, the underlying causes of poverty and overcrowding must be tackled.

DON'T FORGET

Malaria is by far the most commonly used case study and produces the best answers in exams because there is so much information available.

DISEASE CASE STUDY 3: BILHARZIA

Bilharzia (or **schistosomiasis**) affects 200 million people in over 70 developing countries – one tenth will develop severe symptoms including **kidney failure**, **bladder cancer** and death. It is spread by a **parasitic infection**, carried by **snails** living in water where people swim, fish, work and wash.

Bilharzia can be treated with relatively cheap drugs, the host snails can be killed using a **molluscicide**, and modifications can be made to the areas where they breed. Adequate sanitation systems and public-health programmes can also help to control infection.

BENEFITS OF DISEASE CONTROL

Controlling disease means that less money is spent on medicine and health care and can be used for other purposes, including education and reducing national debt. An increase in tourists and a healthier workforce will increase economic development. Other benefits include longer life expectancy and decreased infant mortality.

Information and exercises on development and health from the BBC can be found at http://www.bbc.co.uk/scotland/education/int/geog/health/index.shtml

LET'S THINK ABOUT THIS

More than half the marks for the Development and Health question may be for describing the factors contributing to the spread of a disease, the methods of control and how effective those methods have been, so take the time to learn your case study well.

SOME FINAL TIPS FOR EXAM SUCCESS

Many of the tips you need to get you through the exam successfully are contained in the **Don't forget** and **Let's think about this** boxes throughout the book. A few more, and some final reminders, are shown below.

- Read each question carefully and look for words in bold – these are useful prompts to direct you and help you get the answer right. **Describe** means say what something is like, **explain** means say why it is like that.

- Make sure you look at the number of marks on offer for each question – there's no point writing a whole page for a 4-mark answer.

- Don't use reversals by stating the opposite of what the question asks in your answer or reversing your own statements (e.g. "there is a high birth rate due to lack of family planning and a low birth rate due to family planning").

- Examiners will not give full marks for lists and bullet points, so only use them if you are very short of time on your final answer.

- Some questions seem to come up almost every year, but remember it is very difficult to predict which part of a topic you will be questioned on.

- You can still give grid references in Ordnance Survey-based questions even when you're not asked for them – make sure they are six-figure references unless referring to a very large area.

- If you want to add a diagram to an answer, do so. Marks will always be scored for relevant diagrams, even if the question doesn't specifically ask for them.

- If specifically asked for diagrams, you will lose marks if you don't include them.

- Practise drawing diagrams as part of your revision.

- For most questions in paper 2, you will be presented with reference tables, maps or graphs. Some may just be memory prompts, but, if asked to describe the information given, make sure you quote figures directly from the data.

- For the longer questions (especially those in paper 2), it may be worth making a list of the points you want to cover and ticking them off as you answer.

- Case studies are all-important in paper 2, and every Interaction will refer to them – make sure you have learned yours well. Marks are usually awarded for appropriate named examples from within your case studies.

 If you log on to the SQA site www.understandingstandards.org.uk you can mark your own sample exam questions and compare the results with SQA markers.

- Practice makes perfect, so have a go at writing a timed answer at a rate of less than one minute per mark. Work at it, and you will quickly improve.

- Avoid repeating points, making irrelevant comments or writing out the question as part of your answer – this takes up time and won't get you any more marks.

- The examiners always point out that not enough detail is included on named examples. Don't just mention what your case study is then forget about it; refer to it throughout your answer.

- Practise past paper questions – either those given to you by your teachers or by getting hold of the past paper books in the shops.

- Remember that the marks scheme changed to a whole mark for a valid point in 2008. Questions you look at on past papers pre-2008 will have a half-mark for each valid point.

- Most importantly, read the questions carefully and answer the question asked rather than just writing down all the general information you can think of about the topic.

 It's worth having a look at previous years' examiners' reports from the SQA at http://www.sqa.org.uk/sqa/2517.html. These give some hints on common problems students have in answering questions and how to avoid them.

LET'S THINK ABOUT THIS

As with all your exams, preparation before and on the day is crucial to exam success. If you are well prepared, you won't panic and will do well. Good luck!

INDEX